Creating Good Data: A Guide to Dataset Structure and Data Representation

Harry J. Foxwell
Fairfax, VA, USA

ISBN-13 (pbk): 978-1-4842-6102-6
https://doi.org/10.1007/978-1-4842-6103-3

ISBN-13 (electronic): 978-1-4842-6103-3

Managing Director, Apress Media LLC: Welmoed Spahr
Acquisitions Editor: Susan McDermott
Development Editor: Laura Berendson
Coordinating Editor: Jessica Vakili

Distributed to the book trade worldwide by Springer Science+Business Media New York, 1 NY Plaza, New York NY 10004. Phone 1-800-SPRINGER, fax (201) 348-4505, e-mail orders-ny@springer-sbm.com, or visit www.springeronline.com. Apress Media, LLC is a California LLC and the sole member (owner) is Springer Science + Business Media Finance Inc (SSBM Finance Inc). SSBM Finance Inc is a **Delaware** corporation.

For information on translations, please e-mail booktranslations@springernature.com; for reprint, paperback, or audio rights, please e-mail bookpermissions@springernature.com.

Apress titles may be purchased in bulk for academic, corporate, or promotional use. eBook versions and licenses are also available for most titles. For more information, reference our Print and eBook Bulk Sales web page at http://www.apress.com/bulk-sales.

Any source code or other supplementary material referenced by the author in this book is available to readers on GitHub via the book's product page, located at www.apress.com/9781484261026. For more detailed information, please visit http://www.apress.com/source-code.

Printed on acid-free paper

Creating Good Data

A Guide to Dataset Structure
and Data Representation

Harry J. Foxwell

Apress®

To Eileen, for her endless love and support.

Table of Contents

About the Author

Dr. Harry J. Foxwell teaches graduate data analytics courses at George Mason University's Department of Information Sciences and Technology. He draws on his decades of prior experience as a Principal System Engineer for Oracle and for other major IT companies to help his students understand the concepts, tools, and practices of big data projects. He is a coauthor of several books on operating systems administration and is a designer of the data analytics curricula for his university courses. He is also a US Army combat veteran, having served in Vietnam as a Platoon Sergeant in the 1st Infantry Division. He lives in Fairfax, Virginia, with his wife Eileen and two bothersome cats. Find out more about him at https://cs.gmu.edu/~hfoxwell/.

About the Technical Reviewer

Thomas Plunkett has extensive experience with big data and data analytics. He has taught university courses on related technical topics.

Acknowledgments

I have benefited greatly from valuable encouragement and support for this work from numerous colleagues at George Mason University. Dr. James Baldo, Director of the Data Analytics Engineering program, provided helpful early advice and focus suggestions. And special thanks to Ms. Vidhyasri Ganapathi, Teaching Assistant for several of my data analytics courses, for identifying students' challenges in learning and practicing data science and for confirming their need for this guidance in preparing good datasets.

Introduction

Extracting actionable knowledge from data is a major ongoing challenge of modern IT in corporations, governments, and academia. Creating effectively usable datasets requires an understanding of data quality issues and of data types and the related analytics which can properly be applied. There are numerous data analytics resources – books, articles, blogs, and even commercial software – describing how to clean up and transform data *after* it has been collected, yet there is little practical guidance on how to avoid or minimize the typical "data cleaning" tasks beforehand. Such guidance and best practices are needed to eliminate or reduce lengthy dataset preparation.

Data analysts are often simply presented with datasets for exploration and study which are poorly designed, leading to difficulties in interpretation and to delays in producing usable results. In fact, some analysts report spending up to 80% of their time just getting data ready to be explored so that it can be effectively interpreted. And much data analytics training and published resources focus on how to clean and transform datasets before serious analyses can even begin. Inappropriate or confusing representations, unit of measurement choices, coding errors, missing values, outliers, and others can be avoided by using *good data item selection*, *good dataset design and collection*, and by understanding how data types determine the kinds of analyses that can be performed.

Why not *create good data* from the start, keeping in mind how it will be used, rather than fixing it *after* it is collected?

Creating Good Data discusses the principles and best practices of dataset *creation* and covers basic data types and their related appropriate statistics and visualizations. Following these guidelines results in more effective analyses and presentations of your research data. A key focus of this book is *why* certain data types and structures are chosen for representing concepts and measurements, in contrast to the usual discussions of *how* to analyze a specific data type once it has been selected.

CHAPTER 1

The Need for Good Data

Without data you're just another person with an opinion.

—W. Edwards Deming, Data Scientist [1]

Learning about data analytics tools and methods typically begins with discussions of how to prepare a given dataset for analysis. The reason for this is that many datasets have problems – defects in design, missing or incorrect data items, and non-standard file formats. This often leads to lengthy and complex tasks required to produce datasets ready for efficient analysis. Unfortunately, the critical first step – understanding the nature of data representation – is frequently missing or not sufficiently addressed in resources about data analytics, especially for practitioners just starting their technical careers. Thus, in this chapter, we start with the detailed understanding of data – what it is, how it is expressed, and what we mean by "good" and "bad" data. Only by basing your analyses on *good data* will you produce trustworthy interpretations of your research, leading to good decisions and knowledge-based actions. Let's get started.

Who This Book Is For

The demand for data analytics professionals is growing dramatically. Universities are scrambling to train new analysts and scientists, and this is reflected in the number of new courses, books, and other resources which focus on tools and methods for extracting knowledge from data. *Creating Good Data* focuses on the starting point for analysis – data creation – for those whose tasks include gathering and interpreting data from any discipline:

- Industry, business, and academic researchers and practitioners – anyone who makes decisions based on data analytics

© Harry J. Foxwell 2020
H. J. Foxwell, *Creating Good Data*, https://doi.org/10.1007/978-1-4842-6103-3_1

- New data analysts and data scientists starting their careers

- Corporate trainers and university instructors who teach data analytics

- Students who are learning methods and tools for exploring data

Assumptions

We assume you have a basic knowledge of statistical methods and tools for summarizing and visualizing datasets, including using tools such as R, Python, and SQL, and perhaps some familiarity with commercial software such as SAS, SPSS, and Tableau. Many of you likely already have a library of data analytics texts and other resources that cover data cleaning and presentation, but who would like "early intervention" in dataset design.

All professionals in the rapidly growing data analytics field can benefit from instruction on creating data themselves or on guiding others who will create datasets for their analyses. Data analysts who are called upon to explore and explain other researchers' data can thus guide and encourage the creation of better datasets.

Readers of *Creating Good Data* will use it regularly as a reference, for practitioners as well as for students taking data analytics courses. The book can also serve as a supplementary textbook for such courses.

By the end of *Creating Good Data*, you will understand

- Principles and best practices for creating and collecting data

- Basic data types and representations

- How to select data types, anticipating analysis goals

- Dataset formats and best practices for creating and sharing datasets

- Examples and use cases (good and bad)

- Dataset creation and cleaning tools

And you will be able to create datasets that

- Clearly represent the measurements, quantities, and characteristics relevant to your research

- Minimize time-consuming data cleaning prior to analysis

- Permit clear and accurate statistical summaries and visualizations

Brief code examples from R, Python, and SQL will be included, but this book is not intended to be a complete tutorial for data analysis coding in those languages – there are plenty of those [2,3,4]. Our focus will be on dataset *format* and data *representation* using those programming tools.

The Importance of Getting Data Right

Research and exploration of any kind frequently starts with an idea, inspiration, or curious observation about some phenomenon. Then some claim is made about the nature of that phenomenon. Data provides *evidence* for or against the claim. Without evidence – good evidence (i.e., good data!) – such claims are essentially worthless. And we approach the process of validating or falsifying the claim with a *scientific attitude* [5]. That is, we *care about evidence* and will *change our assumptions and theories* if new evidence requires such change. That's why the field of *data analytics* (the "synthesis of knowledge from information") is part of *data science*:

> *…the extraction of useful knowledge directly from data through a process of discovery, or of hypothesis formulation and hypothesis testing. [6]*

A *data scientist* must therefore understand and implement the concepts, tools, and processes necessary to create, manage, and extract *value* from data, from the creation of the data through to the decisions and actions based upon the analytical results. Figure 1-1 illustrates a typical data analytics process. In this book, we focus on the initial steps needed to produce *good data* and to minimize time-consuming data cleaning and transformation tasks.

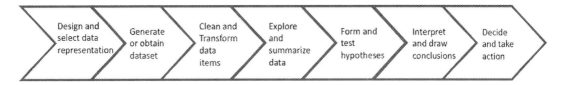

Figure 1-1. *Typical steps in the data analytics process*

What Exactly Is "Data" and Where Does It Come From?

Informally, "data" can be thought of as any collection of symbols representing a set of measurements or observations about some event or occurrence. Other meanings might include lists of "facts" or "statistics," although any collection of words, documents, web pages, and emails can also be considered data. Some such data is purposely designed and collected, as in scientific studies, but other data might be considered "accidental" – likely no one purposely designed Twitter as a formal data collection system, yet today it has evolved into a rich mine of useful knowledge about political and social sentiment, and even a source of information about public health and disease epidemic outbreaks.

More specifically, and we will say more about this in the next chapters, data consists of numbers, characters, words, images, and other symbols, which have definitive *types and characteristics* that directly imply how to summarize and visualize their meanings and relationships.

Our interconnected, digital world is awash with *digital data*. Social media, commerce and business records, scientific measurements, sports statistics, government records, traffic surveillance, health records, wearable devices – the list is endless. The sheer amount of that data and the speed with which it comes at us is enormous and growing rapidly. For example, in a *single minute* of Internet activity, shown in Figure 1-2, nearly half a million Tweets, a million Facebook logins, almost two million emails, and 18 million text messages are happening, and that's just mostly from social media and from personal and business communications.

Figure 1-2. *Data generated during a single Internet minute in 2018 [7]*
`www.visualcapitalist.com/internet-minute-2018/`

What Is "Good" Data?

Good data comes from explicit design and collection decisions about how to represent individual data items and how to present them in a dataset. It permits timely, informative, and ethical analytics and conclusions. Good data items have several critical characteristics needed to ensure valid and useful analysis:

- Accuracy
    - Measurements and characteristics must correctly reflect what is being observed.

5

- Relevance

  - Items selected for analysis must directly relate to the phenomenon being studied.

- Representative

  - Data types must be chosen appropriately to reflect what is being studied.

- Well-defined

  - Data items' meanings must be unambiguously defined in a schema, metadata, or data dictionary.

- Complete

  - Selected data items must include all potentially relevant measurements and characteristics.

- Granular

  - Selected data types should have sufficient range and detail to capture the full variability of the data items.

"Data are people" [8]. Getting data "right" can have important, at times life-critical, consequences – like data from testing the effectiveness of the Ebola vaccine, calculating consumer financial decisions based on credit scores, or determining sentences for crimes. Awareness and ethical practices concerning human-relevant data should always be implemented in data selection and dataset management.

Good data even has the potential for changing fundamental beliefs. The astronomer Kepler was taught and strongly believed that planetary orbits must be perfect circles; his Mars data proved otherwise and led to his famous formulation of the laws of motion for the planets. And today, climate scientists produce and publish data with the hope of convincing the world about the dangers of climate change. Bad climate study data and analysis simply encourages dangerous climate change denial; good climate data has the potential for changing minds.

Where "Bad" Data Comes From

Bad data, on the other hand, hinders or delays analysis and almost certainly results in misleading, inaccurate, or even harmful conclusions. The well-known phrase "garbage in, garbage out" succinctly describes this situation. The sources of bad data include pre-collection design decisions, collection errors, and post-collection interpretation errors. Understanding these sources and planning to address them is essential to effective and accurate dataset analysis.

Some Causes of Bad Data

"Bad" data is often the result of *human error* and poor planning. Obtaining good data is a complex process with many opportunities for mistakes:

- Creation and pre-collection errors

    - Methodological failures: Poorly designed experiments, surveys, or instrumentation

    - Bad documentation: Unclear definitions of terms and missing or confusing schema, metadata, or data dictionary

    - Misspecification of data types and formats: Misunderstanding the purpose and selection of data types and forgetting or avoiding standard data types

- Collection errors

    - Poor collection instructions and methods: Lack of clear processes for recording data

    - Ineffective enforcement of data recording rules: Lack of monitoring and oversight

    - Misinterpretation of data items: Lack of clarity about item meanings

    - Transcription/typos: No checking or validation of recorded data

    - Fraudulent answers/observations: Purposely misleading or nonsensical responses

    - Missing data: Failure to understand and correct reasons for non-answers

    - Impossible, out-of-range data: No bounds checking on data items

- Post-collection and analysis errors

    - Moving/copying: Recording or storage-related mistakes

    - Misinterpretation: Misunderstanding meaning of data items or responses

    - Timeliness, data "rot": Expiration of times, locations, or other characteristics

- Some typical recording errors

    - Recording numbers with leading zeros (0013 instead of 13)

    - Using uppercase letter O for number zero (0); hard to spot

    - Transposing letters or numbers (LA for AL, 32 for 23)

    - Inconsistent use of naming conventions (Italy/Italia, US/USA, Germany/Deutchland/Allemagne)

Preventive Action

Some data analysts report spending the *majority of their time* on a project cleaning, transforming, and preparing their assigned datasets [9]. Obviously, this is costly in time, money, and technical resources. And as Deming also points out, trying to solve this problem *after* the data have been created is not an effective solution:

> *Inspection does not improve the quality, nor guarantee quality. Inspection is too late. The quality, good or bad, is already in the product. —W. Edwards Deming [1]*

That is, the "product" – data – needs to be *created from the start* using practices and components that at least minimize the "bugs" in your datasets. Of course, eliminating all such problems is probably not possible, but if you can get off to a good start with your analytical projects' data, you will produce better and more trusted results.

Summary

In this introductory chapter, we learned about the need for good data, what we mean by "good" and "bad" data, and the origins of potential dataset problems. Minimizing such problems requires awareness of how data collection can fail and by using procedures that ensure quality project design and execution.

The next chapter examines the numerous data types and formats which can be used to represent observations. Thoroughly understanding these data characteristics and using them appropriately will help you significantly in designing and executing your research.

Chapter References

[1] *W. Edwards Deming Quotes*, `https://quotes.deming.org/`

[2] Mailund, Thomas. *Beginning Data Science in R*. New York NY: Apress, 2017.

[3] Hui, Eric. *Learn R for Applied Statistics*. New York NY: Apress, 2019.

[4] Nelli, Fabio. *Python Data Analytics*. New York NY: Apress, 2018.

[5] McIntyre, Lee. *The Scientific Attitude*. Cambridge MA: MIT Press, 2019.

[6] NIST, *Special Publication 1500-1: Big Data Interoperability Framework: Volume 1, Definitions*, 2017, `https://bigdatawg.nist.gov/_uploadfiles/NIST.SP.1500-1r1.pdf`

[7] Desjardins, Jeff; *"What Happens in an Internet Minute in 2018?,"* Visual Capitalist, May 14, 2018, `www.visualcapitalist.com/internet-minute-2018/`, used by permission.

[8] *Ten simple rules for responsible big data research*, `https://dash.harvard.edu/bitstream/handle/1/32630692/5373508.pdf`

[9] *Only 3% of Companies' Data Meets Basic Quality Standards*, `https://hbr.org/2017/09/only-3-of-companies-data-meets-basic-quality-standards`

CHAPTER 2

Basic Data Types and When to Use Them

Without a systematic way to start and keep data clean, bad data will happen.

—Donato Diorio [1]

Decisions about how to represent data measurements for your research projects have important consequences – they directly determine what kinds of statistical and visualization methods can ultimately be used for analysis and presentation of your results. This means you need to select representation types thoughtfully with your analytical goals in mind while at the same time trying to avoid any form of bias in what you decide to measure and what you anticipate your data exploration and analysis tasks will look like.

When we consider a "type" for a data item, we need to specify its context and purpose for our discussion. For programming languages, we define *computational data types* (storage formats), such as *integer* (short and long), *floating point* (single and double precision), *character* (single and multicharacter strings), *Boolean* (0/1, T/F), and derived types such as pointers, how many bits or bytes they use, and how they are referenced by the syntax of the language.

For data analytics, however, we focus on *how* a data item is to be used and interpreted and so refer to *analytical data types*. Additionally, we categorize such data items as *qualitative* or *quantitative*, and then we discuss how varieties within these two categories represent specific measurement requirements.

H. J. Foxwell, *Creating Good Data*, https://doi.org/10.1007/978-1-4842-6103-3_2

Numeric quantities and non-numeric measurements and characteristics can also be classified into several subtypes depending on their levels of descriptive precision and granularity. In this chapter, we discuss the varieties of *analytical data types* and review their typical statistical and visual summary methods.

Four Analytic Data Types

Keep in mind that the purpose of data analytics is to *describe*, *compare*, and *predict* useful insights about some phenomenon. Correctly choosing data types for those tasks must therefore anticipate and enhance our ability to obtain those insights, and such choices ultimately will determine the proper and relevant tools and methods that can be used in our analyses.

Before you can describe a phenomenon and select a representation for it, you must first *understand* and *define* it. Even for something as clear and "obvious" as a physical quantity (e.g., mass) or as elusive as a mental attitude (e.g., prejudice), *precise definition* is needed. This helps to avoid misinterpretation during later analysis. So, for example, you might specifically define a survey respondent's *age* as "the exact number of years and months since their officially recorded birthday" or define their *politics* as "their most recent registered membership in a recognized political party." Only after properly defining what you need to characterize can you select an appropriate representation type. Such definitions must have the potential to capture the full range of possible values for what is being measured or characterized.

Note   There is no absolute correct way to represent a quality or quantity, only what is helpful in description and comparison for your analytic requirements.

We will now discuss the four generally accepted data types, Nominal, Ordinal, Interval, and Ratio (often abbreviated NOIR), and include brief Python or R code examples to illustrate basic statistical and visualization methods for the data types presented.

Nominal/Categorical Data

Nominal (also called categorical) measures are used to capture *qualitative* attributes that have no size or extent characteristics. They are simply *labels* or *names* for some observed attribute such as country of birth, occupation, manufacturing brand, or language spoken. This also includes *binomial* (or dichotomous) characteristics such as true/false, yes/no, or agree/disagree. Such measures explicitly have no implication of order among the labels. For example, for a person's primary spoken language (English, French, Spanish, Farsi, Chinese, etc.), there is no implied order among the languages; you can't claim that French is "bigger" than English in any linguistic or mathematical sense (unless perhaps if you are from France!). Moreover, you can't do any kind of arithmetical operations among the category members – there is no concept of a "mean language."

A nominal data representation must have two important characteristics: its categories must be *exhaustive* and *mutually exclusive*. Exhaustive means that the categories cover *all possible values* in some manner, although there might be a generic "other" category that encompasses multiple cases of low frequency or importance. Mutually exclusive means there cannot be any cases that belong to more than one category. Nominal data items might have many categories represented (e.g., *country of birth*, where there are nearly 200) or only a few such as *gender* (male or female).

Tip   Many studies have used *gender* as a qualifying nominal variable, but such classification is not always well-defined and recent usage might include "other" or specific "non-binary" values. Be aware of any such relevant ambiguities in the data items you select for your analysis.

Because there is no implied quantity for a nominal data item, this limits the kinds of summary statistics, visualizations, and comparisons that are allowable for analysis. Nominal data item collections don't have means, maxima or minima, or measures of variability like standard deviations. All that is possible to characterize such data are *frequencies of occurrence* – how many there are in each category, which can be expressed as absolute counts or as percentages or proportions of the total. Visual summaries of such data include various forms of bar charts and pie* charts. And when counting the frequency of items in each possible category, the category with the largest number of items is the *mode* (although there might be several categories with relatively high frequencies, referred to as *multimodal*).

To illustrate *nominal* data (and subsequent data types), let's examine a sample *synthetic dataset* (constructed for illustration) of hypothetical college graduates, GD-Data.csv [3]. Only the first ten records of 500 are shown:

```
gender;age;degree;field;wrkfld;annsal;payfair;jobsat
Female;40;BS;Engr;Yes;78.0;4;4
Male;39;MS;Engr;Yes;64.0;4;4
Male;36;MS;Comp;No;70.0;3;4
Male;42;MS;Comp;Yes;85.0;5;3
NotSay;39;BS;Comp;Yes;71.0;5;3
Male;38;MS;Biol;Yes;113.0;3;4
Male;38;MS;Comp;Yes;84.0;3;3
Male;37;MS;Chem;Yes;61.0;3;2
Other;28;MS;Chem;Yes;72.5;3;2
Female;31;MS;Comp;Yes;73.5;4;3
...
```

Data fields and types in this dataset are defined as follows:

```
gender: Male, Female, Other, NotSay (String/Nominal)
age: years old (Integer/Ratio)
degree: BS, MS, PHD (String/Ordinal)
field: Engr, Comp, Biol, Phys, Chem (String/Nominal)
workfield: (working in field?) Yes, No (String/Nominal)
annsal: annual salary in $K (Float/Ratio)
payfair: (are paid fairly?) 1-5 (Integer/Likert)
jobsat: (are satisfied in job?) 1-5 (Integer/Likert)
...
```

As we see from Listing 2-1, field is a *character string* representing a *category*, so the permissible analytics for this data item includes counts, relative frequencies, and bar chart visualizations, as shown in Figure 2-1.

Listing 2-1. Descriptive statistics for nominal data item `field`

```
In[37]: data.field.value_counts()
Out[37]:
Biol    160
Comp    144
Chem    120
Engr     46
Phys     28
Name: field, dtype: int64

In[38]: round(data.field.value_counts(normalize=True),2)
Out[38]:
Biol    0.32
Comp    0.29
Chem    0.24
Engr    0.09
Phys    0.06
Name: field, dtype: float64
```

We can also produce this basic analysis using Python along with the *pandas* library (Listing 2-2).

Listing 2-2. Sample Python code for a nominal data item `field`

```
import pandas as pd
data = pd.read_csv('.../Good_Data/GD-Data.csv', sep=';')
data
data.field.describe()
data.field.value_counts()
round(data.field.value_counts(normalize=True),2)
data.field.value_counts().plot(kind='bar')
```

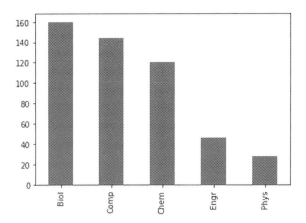

Figure 2-1. *Bar chart visualization for nominal data item* field

Note that we see the counts for each field and easily see the *mode* (most common field). As a general rule, use *nominal data item methods* when analyzing non-numeric characteristics (qualities) that have no intrinsic order.

Hint   If you have a large number of classifications represented in a set of data items, consider grouping very low-frequency items into one category, such as "other."

Ordinal Data

A special case of nominal/categorical data is *ordinal*, where there is an *implied quantitative difference* among the categories but not so specific as to allow for mathematical comparison. In our example dataset, degree clearly has some quantitative characteristic – the "amount" of educational training acquired. But there is no mathematical interpretation of the relative differences in those amounts as specified.

Just like nominal data, summaries of ordinal data types use counts, frequencies, percentages, and proportions, along with bar charts. You can use the same methods. For example, Listings 2-3 and 2-4 and Figure 2-2 show statistics and bar chart visualization for ordinal data item degree.

Listing 2-3. Sample Python code for a nominal data item degree

```
data.degree.describe()
data.degree.value_counts()
round(data.degree.value_counts(normalize=True),2)
data.degree.value_counts().plot(kind='bar')
```

Listing 2-4. Descriptive statistics for ordinal data item degree

```
In[52]: data.degree.value_counts()
Out[52]:
MS      286
BS      161
PHD      51
Name: degree, dtype: int64

In[53]: round(data.degree.value_counts(normalize=True),2)
Out[53]:
MS      0.57
BS      0.32
PHD     0.10
Name: degree, dtype: float64
```

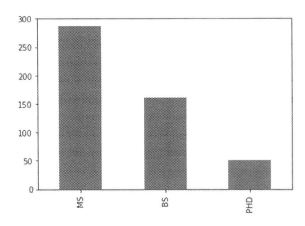

Figure 2-2. Bar chart visualization for ordinal data item degree

Interestingly, depending on the domain and context of your study, what appears to be a nominal data item could be interpreted as an ordinal data item. Take color, for example. In some cases, only the color *quality* is interpreted (e.g., hair color: red, blond, brown, black, etc.), but as a physical quantity/measurement, colors *can* have an implied order: *red > yellow > blue* with respect to their wavelengths. Such distinctions arise from the intended use of the data item and how it relates to other items in the dataset. As we will see later in this chapter, *calendar years* can be interpreted as nominal or ordinal data (and even interval or ratio under certain circumstances).

Special Ordinal Data Types

Certain ordinal data types can be a bit tricky to characterize and to select appropriate summarizations for. Two such related ordinal types are *ranks* and *Likert* (pronounced "Lick-ert") *scales*.

Ranks are numerical assignments of order, as in *class rank* (e.g., graduating number 3 in a ranked class of 150). And like other ordinal data, there is no interpretation of magnitude for a rank; that is, a rank value of 15 is not "5 times worse" than a rank value of 3. In some instances, however, there might be value in reporting "average rank" for comparisons of subsets of two or more ranked groups. One of the most common calculations for ranked data is the ranked item's *change over time*, for example, a university's prestige ranking increasing from tenth to fifth from one year to the next.

Rank scores can give rise to *percentiles* – a calculation of the proportion of a population that fall below a specific value (or rank). For example, a child whose height is reported as *at the 35th percentile* means that her height is greater than or equal to 35% of others in her cohort. An important point here is that certain typical summary statistics such as the arithmetic mean are not valid for all data types; averaging medians or percentiles is not mathematically valid since it hides the effect of extreme values. To illustrate, the medians of the sets {50, 100, 150} and {50, 100, 1000} are the same.

The code in Listings 2-5 and 2-6 shows percentile calculations for annual salaries (annsal) in the example dataset [3].

Listing 2-5. Sample Python code for computing percentiles for ratio data item annsal

```
data.annsal.quantile(.10)
data.annsal.quantile(.25)
data.annsal.quantile(.50)   # the median
data.annsal.quantile(.75)
data.annsal.quantile(.90)
```

Listing 2-6. Percentile statistics for ratio data item `annsal`

```
[In[61]: data.annsal.quantile(.10)
Out[61]: 64.85

In[62]: data.annsal.quantile(.25)
Out[62]: 73.05

In[63]: data.annsal.quantile(.50)   # the median
Out[63]: 89.0

In[64]: data.annsal.quantile(.75)
Out[64]: 101.5

In[65]: data.annsal.quantile(.90)
Out[65]: 110.5
```

Tip   Choose only data types for which you understand the limitations of their allowable summary and relationship statistics.

Likert-scale data is typical of survey responses to questions about a proposed statement; it's intended to capture a *range* of respondents' attitudes or beliefs, usually using an *ordered scale* of five or seven ranked elements (occasionally as many as nine) about a level of *importance, agreement, quality,* or *probability.* For example, a question might ask:

```
"How likely are you to recommend this book to other readers?"

[ ] 1: Very unlikely
[ ] 2: Somewhat unlikely
[ ] 3: Uncertain
[ ] 4: Somewhat likely
[ ] 5: Very likely
```

As with this example, most such Likert-form questions are somewhat symmetric around a middle "neutral" point. And although some statisticians caution against treating such data as anything but *ordinal,* in practice most survey researchers treat it like *interval* data, allowing for computing an *arithmetic mean* value (along with

19

the usual *median* and *mode* statistics) and even *standard deviation*. Generally, such calculations are justified when there is sufficient granularity in the scale and enough responses to ensure roughly normal distribution of the values. Of course, basic frequencies and bar chart visualizations are still appropriate for this type of essentially *ordinal* data. Listing 2-7 and Figure 2-3 show summary statistics and bar chart for Likert data item `payfair` from the sample dataset using the following Python code:

```
data.payfair.value_counts(sort=False)
round(data.payfair.value_counts(sort=False, normalize=True),2)
data.payfair.describe()
data.payfair.value_counts(sort=False).plot(kind='bar')
```

Listing 2-7. Summary statistics for Likert data item `payfair`

```
In[90]: data.payfair.value_counts(sort=False)
Out[90]:
1     58
2     121
3     111
4     127
5     81
Name: payfair, dtype: int64

In[91]: round(data.payfair.value_counts(sort=False, normalize=True),2)
Out[91]:
1     0.12
2     0.24
3     0.22
4     0.26
5     0.16
Name: payfair, dtype: float64

In[92]: data.payfair.describe()
Out[92]:
count     498.000000
mean       3.104418
std        1.267589
min        1.000000
```

```
25%          2.000000
50%          3.000000
75%          4.000000
max          5.000000
Name: payfair, dtype: float64
```

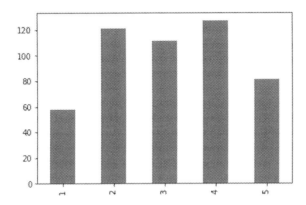

Figure 2-3. *Bar chart for Likert data item* `payfair`

Ratio Data

Ratio data typically represents physically measurable *quantities* that have a possibility of being *zero* (none). Counts of items, distances, durations, geometric measurements (length, height, width, area, volume), weight and mass, and costs and prices are all examples of ratio data. Allowable calculations on such data include means, medians, differences, and of course ratios (thus the name). In the sample dataset `GD-Data.csv`, both `age` and `annsal` are examples of ratio data. That is, differences are interpretable (25 years old is 5 years less than 30 years old), ratios are interpretable (30 years old is twice 15 years old), and an `age` of *zero* has meaning. Statistical summaries and visualizations of such ratio data include means, ranges, boxplots, and histograms, as shown in Listings 2-8 and 2-9 and Figure 2-4. Boxplots are particularly effective for visualizing the range (min and max values), quartiles, and median:

```
data.annsal.describe()
data.boxplot('annsal')
data.hist('annsal')
```

Listing 2-8. Summary statistics for ratio data item `annsal`

```
In[38]: data.annsal.describe()

Out[38]:
count    498.000000
mean      87.780522
std       16.756242
min       55.000000
25%       73.050000
50%       89.000000
75%      101.500000
max      122.700000
Name: annsal, dtype: float64
```

Listing 2-9. Boxplot for ratio data item `annsal`

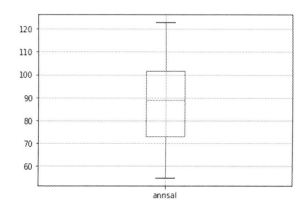

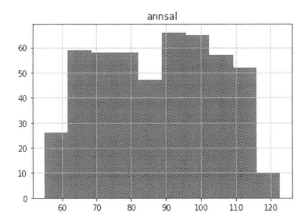

Figure 2-4. *Histogram for ratio data item* `annsal`

Ratio data consists of quantitative measurements, and there is a great variety of how such numbers can be represented. We examine that variety in more detail in Chapter 3.

Interval Data

Interval data can be considered as a special case of ratio data, with the exception that *ratio calculations are not interpretable* since there is no absolute zero quantity defined. The typical example for such data is temperature measured in degrees Fahrenheit or Celsius, both of which have a "zero" value, but which do not indicate absence of temperature. To clarify this, consider a reported daytime high temperature of 90° F, along with a reported daytime low of 60° F. Multiple temperature measurements throughout the day could be characterized using a *mean* temperature for the day, there is an interpretable difference of 30° between the high and low temperatures, but it is *not correct* to say that the daytime high is 1.5 times higher than the low. Other examples of interval data include calendar years (e.g., 1492, 1776, 2019), latitude/longitude locations, and even IQ scores. Other than excluding specific ratio relationships, you can generally use the same statistical and visualization methods for interval data as for ratio data (see Table 2-1).

Table 2-1. *Summary of univariate data types*

Analytical Data Type	Computational Data Type	Typical Statistics	Typical Visualizations
Nominal, categorical	Character, string,	Counts, relative frequencies (%), modes	Bar chart, pie chart*
Ordinal	Character, string	Counts, relative frequencies (%), modes, percentiles, quartiles	Bar chart
Interval	Integer, floating point	Max, min, range, mean (arithmetic), median, mode, differences, std dev, addition, subtraction	Boxplot, histogram
Ratio	Integer, floating point	Max, min, range, median, mean (arithmetic and geometric), ratios, differences, std dev, addition, subtraction, multiplication, division	Boxplot, histogram

* Warning   *Don't use pie charts! They are difficult to visually interpret the differences in areas and are susceptible to artistic excess, superfluous dimensionality (3D), and chartjunk (distracting graphical embellishment). Use bar charts instead! If you must pie, use only a few (4–6) large segments, and don't display tiny slices! [4]*

Other Data Types

There are indeed types of data which don't fall neatly into the four described earlier. One of the most common is *text* data, which comes from many sources: books, documents, blogs, emails, social media, web pages, and recordings of spoken language, to name just a few examples. We can think of the four basic data types as "atoms" that make up the "compounds" of complex data objects. So, for text, such data is a collection of nominal/categorical data – words, parts of speech, phrases, sentences, and so on. In fact, a *word* is precisely a *category* for members of a linguistic concept, like "fish" or "book."

Analytical techniques for understanding text include word counts and frequencies, exactly what we typically do with other categorical data. One common use for such text exploration is "sentiment analysis," classifying and measuring the *frequencies* of word categories associated with positive or negative feelings, for example, about a discussion topic, political event, or a product review.

Data analytics projects often include gathering text data in some form, like explaining "other" content responses, giving reasons for answer choices, or providing more detail about a selection. Specialized libraries for analyzing text data, like the Python Natural Language Toolkit (NLTK) [5], are available for exploring and summarizing collections of words.

And what about *media* data like images, videos, music, and spoken language recordings? While these objects clearly don't fall under the four NOIR types, we can still explicitly create descriptions (metadata) which use those types for exploring the content and context of the media. For example, a video object can be described as having duration (ratio data), object presence and count (nominal), and recording type/medium (nominal). So, when considering the descriptions of non-standard data, use the basic types and methods we've just reviewed.

Summary

Understanding the uses and limitations of different data types is critical to correctly creating, exploring, and analyzing datasets. The four NOIR types and associated analysis methods we have covered in this chapter are widely accepted and used in statistical and visualization studies. When you plan your own projects (see Chapter 4), using the right data representations will make life much easier for your data analytics janitorial staff.

Tip   Think of the *analytical data types* we have discussed in this chapter as an *ordinal* data type itself: *nominal < ordinal < interval < ratio*, where specificity and granularity of measurement increases from basic qualitative categories through numerically precise quantities. This will help you recognize and choose the optimal data types for your projects.

We expand a bit further on qualitative/numeric data in the next chapter, cautioning analysts about misuse of units, size characteristics, and related measurement and representation issues.

Chapter References

[1] Lehr, S., *20 Inspirational Quotes About Data*, 2019, www.ringlead.com/blog/20-inspirational-quotes-about-data/

[2] Boone, H. and Boone, D., *Analyzing Likert Data*, Journal of Extension, Vol. 50, No. 2, 2012, www.joe.org/joe/2012april/tt2.php

[3] GD-Data.csv; see GIT repository, https://github.com/hjfphd/CreatingGoodData.

[4] Few, S., *Save the Pies for Dessert*, Visual Business Intelligence Newsletter, 2007, www.perceptualedge.com/articles/visual_business_intelligence/save_the_pies_for_dessert.pdf

[4] *Natural Language Toolkit*, www.nltk.org/

CHAPTER 3

Representing Quantitative Data

One accurate measurement is worth a thousand expert opinions.

—Grace Hopper [1]

In Chapter 2, we learned about four basic analytical data types. Quantitative data – *interval* and *ratio* – are among the most common types and are often the most problematic due to the great variety of measurement units and size ranges. Units vary by knowledge domain, usage conventions, national and historical origins, and formal or organizational standards. Sizes range from the infinitesimal to the cosmic in fields like economics, computer science, quantum physics, and astronomy. Ensuring good quantitative data requires selection and recording of appropriate units and defining them in the data dictionary for the dataset and avoiding any unit labels in the data itself. In this chapter, we review how common quantities and measurements are expressed and caution how mixing unit types can cause time-consuming data cleanup tasks.

Units of Measurement

William Thompson, also known as Lord Kelvin, famously said "When you can measure what you're talking about, and express it in numbers, you know something about it." [2] The problem, then, is *how* to express such measurement. Do you measure distance by the length of the current king's foot or by something more precise and consistent? [3] Today, of course, we have at least some widely accepted and precisely defined standards

© Harry J. Foxwell 2020
H. J. Foxwell, *Creating Good Data*, https://doi.org/10.1007/978-1-4842-6103-3_3

for measuring physical characteristics such as length, mass, and time: the International System of Units (SI), or "Metric System" [4,5], which originated in 1799 and has been updated a number of times, most recently in 2019.

Unfortunately, some researchers don't always use them or misuse them in recording their data. And there are quantities that don't fit exactly into that standard.

Magnitudes and Quantities

How we typically *discuss* very large numbers differs from how we should *represent* them in datasets. For example, familiar computing-related quantities such as throughput, memory sizes, and storage capacities are generally described using *mixed* units: megabits and gigabits per second (mbps, gbps), megabytes and gigabytes (MB, GB) of RAM, and gigabytes, terabytes, and larger (GB, TB) for disk space. A dataset listing sizes of digital archives would probably include recorded values such as "22MB", "13GB", and "6.5TB".

But such representations in a single dataset can make aggregation statistics (e.g., arithmetic means) harder to compute, unless such storage values are expressed uniformly in common/standard units. If not, it's up to the poor data analyst to resolve! So, how can you avoid having to post-process your mixed unit representations? First, *do not include units of measurement in your dataset*. That is, do not record measurements such as "15.3° C" or "3.2m"; simply record "15.3" or "3.2". List the numeric values only and specify the data items' units of measurement in the metadata document.

Tip   *Always* use "naked numbers" in your datasets, only values, and no non-numeric characters representing units of measurement. Indicate units only in your dataset's dictionary, and *never* mix different units in the same set of data items. Use accepted standard units for your specific domain of inquiry whenever possible.

One generally accepted data representation practice is to use scientific notation for measurements with extreme ranges of size. In this format, 14GB is therefore 1.4×10^9 bytes, or preferably 1.4E9. For very small quantities, such as 2 nanoseconds (nsec) or 0.000000002 seconds, record 2.0×10^{-9}, or 2.0E-9. Then, the coefficients can be recorded separately from the exponents in the dataset (labeled of course in the data dictionary), thus avoiding any non-numeric characters attached to the data item (the cause of much data cleaning problems). Better yet, as we show later in this chapter, ensure the data item is recorded in standard scientific notation (#.#E##), and then during subsequent analysis, programming tools and libraries (like Python) can be used which know how to interpret such formats.

Astronomical distances are prime examples of data items needing consistent units, from the mere millions of miles within the solar system to the thousands of trillions of miles for intergalactic distances. World economic and population data can also span many orders of magnitude, and subatomic distances and masses have vanishingly small measurements [6]. Therefore, choose representations that can capture the full range of possible values for such quantities, and consider using scientific notation to express them.

Another general rule for data recording is to prefer *raw* measurements to *derived* values (which can be calculated later as needed during analysis). That is, instead of recording a proportion or percent, record the count and the base in separate data items; this allows for flexibility in preparing statistical summaries, relationships to other items, and more detailed visualizations.

Hint   It should go without saying that you should avoid using "silly" units of measurement, like "how many swimming pools of water flow over Niagara Falls per second" or "how many stacked dollar bills would reach the moon." These are the statistical equivalent of *chartjunk* (unnecessary artistic decorations of graphical information) and have no place in serious data analytics.

More advice: when collecting data, avoid recording it using compound measurement representations, such as 2h14m, 5ft6in, 3lb6oz, and the like. Convert it prior to recording, or your data analyst will need to figure out how to do that for you, adding time and process to obtaining results.

Time Data

What time is it now? When did an event occur? How long did it last? When will it happen again? These are time-based questions whose typical representations can vary widely in format. Recording such data is fraught with potential errors, resulting in time-consuming (no pun intended!) post-processing to prepare for usable summary statistics, comparisons, and visualizations.

Consider how calendar dates are reported, for example, Queen Elizabeth's birthday:

(a) `April 21, 1926`

(b) `21 APR 1926`

(c) `4/21/26`

(d) `21/4/26`

(e) `19260421`

All of these are common date formats; only the last one (e) allows for correct temporal sorting of such expressions; the others might require modification or additional software functions before ordering. And note that (c) and (d) call out the same date, but month first is the *de facto* standard in the United States (ever the oddity!), while day first is common in Europe. If you are recording dates in your dataset, it's best to choose a standard format that permits ordering without the need for modification. Some data analysis software, notably Excel, can adapt to multiple date formats, but it's best to pick a consistent format that is optimal for your analytical tools and requirements.

Clock-time formats are no better in their variety of representation. If we say it's `8:15`, do we mean AM or PM? In what time zone? Is it daylight savings time? Again, if you want to accurately record the exact time of an event, you need some reasonably accepted standard, for example, Coordinated Universal Time (UTC) [7]. Using this format, record times as `HH:MM:SS`, as in `15:41:56`; note in the metadata that this is UTC time, and use analytical tools that can interpret such time formats.

There are multiple ways to report durations, depending on the context and domain of your study. You might record an event lasting 24 seconds, 13 minutes, or 3.2 milliseconds. Whatever you choose, ensure consistency in the units, converting all the data items to the same unit [11], or again you will impose upon your data cleanup staff to fix things.

One interesting time measure in computing is UNIX Epoch Time [8], which is the "number of seconds elapsed since 01 January 1970 UTC." For example, the Epoch Time for 10:00 AM EST, Tuesday, November 12, 2019, is 1573570800 seconds. It's a handy measure for recording occurrence times and durations of computer events (obviously on UNIX, OS X, or Linux operating systems). However, like the earlier Y2K date problem, it has an "expiration date" due to the 32-bit size of the integer used to store it: January 19, 2038, after which the world apparently will end. We mention this problem with Y2K and Epoch Time as a reminder that decisions about how to represent and store data can have severe long-term consequences, as when the representation overflows resulting in incorrect values.

In any case, you should select a format for calendar dates, clock times, and durations which encompass the full range of possible measurements in your study and express them consistently using only one unit of measurement.

Money and Currency Data

We live in a global economy and so are aware of the multitude of world currency types: US dollars ($), European euros (€), and Japanese yen (¥) as examples. When recording money values, avoid using multiple currencies with their attached symbols; convert to a common currency before inclusion in the dataset. Such conversions, however, will depend on the nature of the study and might need (or not) to account for exchange rates.

To make matters worse, be aware that an "American billion" (one thousand million: 1,000,000,000) is not the same as a "European billion" (one million million: 1,000,000,000,000). Also, when recording and reporting large numbers with fractional parts, like "3,245.19", some countries use a space in place of the comma and a comma in place of the decimal point: "3 242,19". This can be quite confusing when acquiring data from international sources; clarifying and standardizing such quantities after dataset collection can also add significant time and effort to your analysis, so ensure consistent units and formats at the start of your data project.

Transformations and Indexing

Sometimes it is helpful to transform measurements to reflect a comparison with a standard unit. This can reduce the representational sizes of the numbers and improve understanding of their relationships. For example, Table 3-1 shows average distances of our solar system's planets from the sun indexed relative to the Earth's distance. The actual distances are in millions of miles; Earth's distance is set as the standard – one astronomical unit (AU) – and the other distances are expressed in that unit.

Table 3-1. *Planetary distances indexed to Earth's distance from the sun [15]*

Planet	Distance (Miles)	Distance (AU)
Mercury	33,980,000	0.39
Venus	67,240,000	0.72
Earth	92,960,000	1.00
Mars	141,600,000	1.52
Jupiter	483,800,000	5.20
Saturn	890,800,000	9.58
Uranus	1,784,000,000	19.19
Neptune	2,793,000,000	30.05

Additional methods for reducing the recorded range and magnitude of a data item are *normalization* and *standardization*. These methods, as well as indexing, can be computed from the raw data during analysis or can be the primary expression of the data when it is collected. Standardization, a common practice, involves subtracting the mean of the data item and dividing by the data's standard deviation, producing transformed data with a mean of 0 and a standard deviation of 1.

Normalizing data, rescaling it based on its range so that all transformed values lie between 0 and 1, is often used in multivariate analysis especially in machine learning. Python includes libraries and functions that support these data transformation methods, but again we generally recommend original raw data values to be recorded if at all possible and only applying the transformations afterward during the analysis.

Measurement Standards

The nice thing about standards is that there are so many of them to choose from.

—Andrew Tanenbaum [9]

We have suggested that standardizing the representation of data items in your datasets is an important goal, one that's especially important for sharing your research [10]. Adhering to universally accepted metrics and descriptions is essential to scientific research for interpreting and replicating experiments and other analytic activities. The *International Organization for Standardization* (ISO) [12], consisting of national standards bodies from around the world, publishes numerous documents defining and clarifying the vocabulary of data collection and analytics [13]. In fact, they include formal definitions for the analytic data types we discussed in Chapter 2. All researchers – practitioners, instructors, and students – should be familiar with ISO standards for representing, collecting, and describing data and datasets.

Other Quantitative Measurement Issues

Recorded quantities should be clearly interpretable, unambiguous, and appropriately precise. Excessive precision, duplication of other data items, and formatting can also create cleanup problems.

Numerical Precision

When recording measurements, decide what level of numerical precision is needed for your specific analytical needs. For example, when measuring a person's height in meters, is 1.8 sufficient or do you really need 1.798? In your expected analysis, how significant are the *differences* among data items in your domain of inquiry? What level of precision is needed for accurate comparative analysis (e.g., linear regression)? Deciding on greater precision generally does not cause data cleanup problems, unless there are large precision differences among the items being compared.

Multicollinearity

When selecting data items for your dataset, two or more items can closely measure the same quantity. As a simple example, suppose you recorded a group's approximate heights in centimeters as well as in inches (i.e., take two independent measurements, not a mathematical conversion). Clearly the two measures would be highly correlated and would adversely affect predictive models that included both items. So instead of requiring multicollinearity tests during analysis [14], avoid this problem by recognizing any direct linear relationships among your chosen data items (if possible; multicollinearity is not always obvious until the dataset is explored).

Non-numeric Numbers

Although the preceding sub-title seems like an oxymoron, there are indeed data items in many studies that use numeric codes as categories. The typical example for this is US Postal Service Zip Codes, such as 22030 (for Fairfax, Virginia). Zip Codes are simply numeric labels (nominal data!) for geographic regions and should not be used for any mathematical calculations; there is no "arithmetic mean Zip Code" for a set of regions, for example.

Similarly, numeric tags for record numbers, employee numbers, US Social Security Numbers, telephone area codes, and other identification tags are generally *nominal* data types and should be analyzed as such. It is possible there could be some temporal ordering to such values (e.g., employee ID sequence reflecting date of hire), but still this is at best ordinal data, and no arithmetic is performed on such data.

Recording numbers with leading zeros can cause analytical tools to misinterpret the measurement as a character string instead of a number, requiring cleanup or extra transformation ("0013" instead of "13"). In general, don't use them unless you really do want to interpret the string as categorical data.

Summary

In this and previous chapters, we have emphasized the need for *good data*, described how to represent and record that data, and briefly discussed some of the practices and difficulties involved. This prepares you for the next step – planning your data collection and analysis projects. The goal is to eliminate or at least minimize data representation problems and cleanup efforts and thus reduce the time required to effectively complete your research.

Chapter References

[1] *All Author Quotes*, https://allauthor.com/quotes/101062/

[2] *Inspirational eQuotes*, https://inspirational-e-quotes.com/authors/lord-kelvin

[3] *The Origin Of 'Foot' For Measurement (+ Why 12 Inches?)*, https://sparkfiles.net/foot-whats-special-12-inches

[4] *SI Units,* www.nist.gov/pml/weights-and-measures/metric-si/si-units

[5] *2019 redefinition of the SI base units,* www.bipm.org/utils/common/pdf/SI-statement.pdf

[6] *Mysterious Neutrinos Get New Mass Estimate,* www.scientificamerican.com/article/mysterious-neutrinos-get-new-mass-estimate1/

[7] *Coordinated Universal Time,* www.timeanddate.com/worldclock/timezone/utc

[8] *UNIX Epoch Time,* www.unixtutorial.org/epoch-time

[9] *Famous Quotes & Sayings,* www.quotes.net/quote/17902

[10] *On the importance of data standards in citizen science,* www.cs-eu.net/blog/importance-data-standards-citizen-science

[11] *What does bad data look like?,* https://medium.com/@bertil_hatt/what-does-bad-data-look-like-91dc2a7bcb7a

[12] *International Organization for Standardization,* www.iso.org/home.html

[13] ISO 3534-2:2006(en) Statistics — Vocabulary and symbols, www.iso.org/obp/ui#iso:std:iso:3534:-2:ed-2:v1:en

[14] *Multicollinearity: Definition, Causes, Examples,* www.statisticshowto.datasciencecentral.com/multicollinearity/

[15] *Planetary Fact Sheet - Ratio to Earth Values,* https://nssdc.gsfc.nasa.gov/planetary/factsheet/planet_table_ratio.html

CHAPTER 4

Planning Your Data Collection and Analysis

There is as yet insufficient data for a meaningful answer.

—COSMIC AC [1]

Chapter 2 emphasized that data types are classified according to their purpose, and Chapter 3 cautioned you about the myriad ways that measurements can be represented. Recall that the general purpose of data analysis is to describe some phenomenon, to explore some potentially informative relationships among the data items, and to model and predict the behavior of that phenomenon. And the inception of your inspiration to study it in the first place implies that you have some sense of what might be useful to measure and what relationships might be lurking in the data. This means you need to think carefully about what to measure and how to measure it. And this does not mean that you ignore potential biases or unfounded assumptions about the focus of your research. Anticipating the tools and methods of your analysis does not mean presupposing anything about your as-yet-uncollected data, but it does require some care in choosing data types that enable the analyses you expect to conduct. A key message of this chapter is that *the plan for your future analysis should guide the selection of your data representations.*

Describing, Comparing, and Predicting

Effective analysis begins with good data, that is, sufficient and relevant measurements with few or no errors, no unnecessary artifacts (e.g., explicit units of measurement), and a dataset format that enables easy exploration, visualization, and sharing for replication

H. J. Foxwell, *Creating Good Data*, https://doi.org/10.1007/978-1-4842-6103-3_4

and further research. Anticipating and visualizing your final analysis and potential conclusions will ensure a good start to your research. Clarify what you are looking for and what you will say or do when you have found it (or haven't!). Then define and collect your required data. Sometimes you will simply be validating your assumptions about a relevant measurement; at other times, you will hope for some interesting or unexpected result. Outliers and other data oddities are not necessarily bad data; some of them can reveal profound insights!

> *The most exciting phrase to hear in science, the one that heralds new discoveries, is not 'Eureka!' but 'That's funny...'*

> —Isaac Asimov [2]

Example: Choosing a Data Type

Since choosing data types is such a critical early task in your research, you need to understand and practice this process.

To illustrate this idea of choosing an appropriate data type for a study, consider this theoretical example: assume you are a health researcher with a suspicion that a person's weight and blood pressure are related. Recording blood pressure is fairly standard (two numbers: diastolic and systolic, both *ratio* data types). How should you represent and record weight data, keeping in mind how you might describe and visualize a suspected link? You could obtain information about a person's weight using any of the following questions or methods:

1) "Describe your weight": _____

2) "Are you overweight?":

    a. Yes/No/Unsure (select one)

3) "How much do you weigh?" _____

4) "What is your weight to the nearest pound?" _____

5) (Nurse records patient's weight reading on a scale):

    _____

6) (An electronic scale records and transmits patient's exact weight): ____

Clearly each of these options provides *some* information about the person's weight, although they certainly vary in *data type* as well as in precision, in accuracy, and in the potential for bias. But considering that you probably expect to prepare some sort of model of a potential relationship between blood pressure components (both *ratio* measures) and weight, you should choose option 4 or 5 (also *ratio* data), anticipating a regression analysis. That is, *the expected analysis type determined the required data type*. Such forethought should be used for *all* data type choices.

Plan for Visualizing Your Data and Analysis

Above all else, show the data.

—Edward R. Tufte [3]

Note   You must *show* your data and analysis, not just list statistics and provide explanatory text.

Visualization – the graphical presentation of your data and analysis – is an absolute necessity for communicating your research results and conclusions. You've no doubt heard that "a picture is worth 1000 words"; we say "a thousand words (or data values) deserves a picture." In fact, it's often difficult or nearly impossible to interpret lengthy and complex statistical tables without corresponding visuals. A perfect example of this is the rather famous Anscombe's Quartet [4], four constructed datasets with clearly different data values, but with identical descriptive statistics (means, standard deviations, correlation coefficients, and linear regression lines). You can't distinguish the real differences among the four datasets without graphing them!

You should therefore plan on creating informative visualizations of your data and analyses; what will your hopefully convincing results look like? And just as you should carefully select data types for your research, you should anticipate and select their corresponding visualizations, even perhaps determining a data type to be used based on how you want to display and communicate information about it.

The Purpose and Goal of *Univariate* Descriptive Statistics and Visualizations

Your first analytical task will be to provide basic information about your individual data items.

As we covered in Chapter 2, each of the four major data types has corresponding descriptive statistics and visualizations. As a brief reminder, *qualitative* data (nominal and ordinal) requires counts and frequencies to summarize them, along with bar charts of various kinds for visualizing those statistics. *Quantitative* data (interval and ratio) requires measures of centrality and dispersion (e.g., means, ranges, variance) and related visualizations like boxplots and histograms. When choosing data types for your research project, consider using types that provide compelling graphical representations that communicate the nature of the characteristics and quantities being measured.

The Purpose and Goal of *Multivariate* Relationship Statistics and Visualizations

After describing your data items, look for relationships.

The most interesting and rewarding exploration of data occurs when it results in the discovery or validation of relationships among pairs or groups of data items. Some historic examples of careful examination of data include Kepler's discovery of the laws of planetary motion [5], FitzRoy's discovery of the relationship between barometric pressure and storms [6], and Snow's discovery of the cause of a massive cholera epidemic in London [7]. Each of these critical discoveries came from careful collection, analysis, and *visualization* of their respective datasets and involved the *recognition of patterns* in the data and the formalization and validation of those patterns. Kepler's hypothesis that the planets orbited in perfect circles was rejected by plotting his data. Snow used a simple geographic plot to locate the source of the infections. And FitzRoy's habit of keeping extensive weather observations led him to uncovering critically important storm predictions. Data exploration and understanding can change our view of the universe and can even save lives.

The human visual system evolved to detect patterns; we easily see shapes, temporal changes, and other distinguishing characteristics in text, lists of numbers, and images. But we also "see" spurious patterns where none exist, for example, the infamous "Face on Mars" [8]. The face was an illusion of shadows that resonated with the brain's ability to detect face-like images.

What we seek in data analytics is real, explainable relationships among data items, identifying patterns of change in one or more items (dependent variables) as a result of changes in other data items (independent variables). And again, the selection of appropriate data types for representing such relationships enables their effective presentation and interpretation.

Independent and Dependent Variables

Identifying statistical dependencies is the key to explaining relationships among your data items.

When exploring a potential relationship between two data items, the typical assumption is that observed changes in one of the items are hypothesized to have some observable effect on the value of the other item. When this occurs, we say the two data items are *correlated* and we associate some appropriate measure of the "strength" of that effect. The controlling item is called the *independent variable* in the relationship in that we are "free" to select a value for it, and the resultant item is called the *dependent variable* in that its value "depends" in some way on the independent variable (remembering, of course, that this potential dependence is not necessarily causal). For example, if we measure the cruising speed of an automobile and its instantaneous rate of fuel consumption, we generally see the consumption increase as the speed increases. So, we state that "the fuel consumption rate *depends* upon the speed," the speed and consumption rate are the respective independent and dependent variables, and we expect to model the consumption rate as a function of speed. Sounds easy, but often it's not!

For example, many US high school students take a two-part college entrance test, the SAT [9], consisting of two parts: verbal and mathematical aptitude. Results of the two parts are *highly correlated*, but that does not imply that there is a direct causal relationship between verbal and mathematical ability. Correlation is simply an observed numerical relationship between two data items; you need to further investigate any apparent relationship. For some entertaining reading, check out Tyler Vigen's *Spurious Correlations* page [15].

Correlation is not causation!   While this statement is well known to statisticians and data analysts, nevertheless it is one of the most frequent data interpretation errors and requires continuous reminding and examination when correlations are proposed.

That does raise the question: when do you have evidence for causation? Generally, this requires additional study and theorizing in the specific domain of the research, including additional well-designed controlled experiments [10]. Causality cannot be verified via data correlation; you need to conduct controlled studies that identify and eliminate unrelated variables and isolate the suspected causal link. The history of science has many examples of the efforts to determine true cause-and-effect phenomena, such as identifying the cause of influenza (virus and *not* bacteria) [19].

Among the most commonly used values and methods for exploring and visualizing relationships among data items are the various types of correlation coefficients and different methods of regression analysis. This leads us back to our focus on selecting types of data to use for our research projects. Linear regression, for example, requires both the independent and dependent variables to be numeric interval or ratio data types. When either or both of these items are nominal or ordinal data, different methods are needed, such as logistic regression, which is generally used to predict the value of a dependent nominal data item. And there are analogous calculations of relationship strength ("correlation"), for combinations of quantitative and qualitative data types, such as ANOVA (analysis of variance) and chi-squared tests [11,12]. Many of these statistics and methods require certain characteristics of the data items, for example, normally distributed ratio data values, along with specific calculations for hypothesis tests. The fundamental concept for such hypothesis tests is the comparison of what is expected by the chosen model to the actual results and the probability of observing the difference.

Choosing appropriate data types for potentially related measurements in your research project should be guided by the way you expect to describe and visualize those relationships. Table 4-1 summarizes some of the most common statistical and visualization methods.

Table 4-1. *Summary of basic bivariate summary statistics and visualizations [11,12]*

Dependent Variable			Nominal / Ordinal (Quantitative)	Interval / Ratio (Qualitative)
Qualitative	Ratio		ANOVA, F-tests	Regression, Correlation, Scatterplots, t-tests, F-tests
	Interval			
Quantitative	Ordinal		Frequencies, Bar charts, Crosstabs, Chi-Square tests	ANOVA, Logit Regression
	Nominal			
			Nominal | **Ordinal** | **Interval** | **Ratio**	
			Quantitative | **Qualitative**	
			Independent Variable	

We leave the specifics of statistical tests and calculations to you and your unique analytical needs. Consult the references for guidance on selecting commonly used and specialized methods.

Data Analysis Tools

What programs should I use to analyze my data? Does it matter?

Among aspiring data analytics professionals, questions and heated arguments often break out about what the "best" tools are for exploring, summarizing, and visualizing data. Without taking sides, we advise analysts to become as fluent as possible in *all* the major offerings: Python, R, SQL, and Excel; these are the top four consistently reported in surveys of the profession, such as O'Reilly Media's Annual Data Science Salary Survey [13], and DataCamp's comparison of Python vs. R [14]. Most analysts use a combination of tools and programming languages, depending on the specific tasks of data representation and transformation, statistical summaries and tests, visualization types and quality, library availability, personal and organizational preferences, and ease of use.

Commercial products such as SAS, SPSS, and Tableau are widely used in industry, government, and academia. SAS and SPSS have a long history of use in statistical analysis and include excellent instruction and tutorials on data types, related tests, and visualizations [16,17]. Tableau is particularly useful for visual data exploration and for producing attractive, publication-quality graphics [18]. And because the field of data analytics is growing rapidly, a growing number of general purpose and domain-specific tools are becoming available, both commercially supported and community-developed open source software. Most of these offerings allow for flexibility and freedom in defining and selecting appropriate and relevant data types for your analytics needs.

Hint   Don't focus on becoming deeply expert in only one analytical tool to the exclusion of the others. Beware of the hammer/nail analogy: when the only tool you have is a hammer, all problems start to look like nails. They're not!

Summary

This chapter provided some guidance in planning your data analytics projects, emphasizing how your analytical goals should determine how you select data types and representations. Always keep in mind the purpose of your investigation, the hoped-for results, and the need to share and communicate those results to other interested parties. Be aware of potential biases and unreasonable assumptions, including misinterpretation of apparent relationships among data items. And although it's admittedly difficult to keep up with current and emerging analytical methods and technologies, spend the time reviewing and learning about what's new and useful in this exciting and dynamic profession.

Next, we'll cover an important component of data analytics: how to format your datasets so that they can be easily shared with other investigators for replication and extension of your results.

Chapter References

[1] *The Last Question*, www.multivax.com/last_question.html

[2] *Asimov Brainy Quotes*, www.brainyquote.com/quotes/isaac_asimov_109758

[3] *The Visual Display of Quantitative Information Quotes*, www.goodreads.com/work/quotes/522245

[4] *Anscombe's Quartet*, https://towardsdatascience.com/fables-of-data-science-anscombes-quartet-2c2e1a07fbe6

[5] *Kepler's laws of planetary motion*, www.britannica.com/science/Keplers-laws-of-planetary-motion

[6] *FitzRoy: The Remarkable Story of Darwin's Captain and the Invention of the Weather Forecast*, www.amazon.com/FitzRoy-Remarkable-Darwins-Invention-Forecast/dp/1530086361/

[7] *John Snow and the 1854 Cholera Outbreak*, www.pastmedicalhistory.co.uk/john-snow-and-the-1854-cholera-outbreak/

[8] *Unmasking the Face on Mars*, https://science.nasa.gov/science-news/science-at-nasa/2001/ast24may_1/

[9] *Scholastic Aptitude Test*, https://collegereadiness.collegeboard.org/sat

[10] *Correlation does not equal causation but How exactly do you determine causation?*, www.datasciencecentral.com/profiles/blogs/correlation-does-not-equal-causation-but-how-exactly-do-you/

[11] *The Statistics Tutor's Quick Guide to Commonly Used Statistical Tests*, www.statstutor.ac.uk/resources/uploaded/tutorsquickguidetostatistics.pdf

[12] *Selecting Statistical Techniques for Social Science Data: A Guide for SAS Users,* www.amazon.com/Selecting-Statistical-Techniques-Social-Science/dp/1580251188

[13] *2017 Data Science Salary Survey,* www.oreilly.com/library/view/2017-data-science/9781491997079/

[14] *Choosing Python or R for Data Analysis? An Infographic,* www.datacamp.com/community/tutorials/r-or-python-for-data-analysis

[15] *Spurious Correlations,* www.tylervigen.com/spurious-correlations

[16] *SAS: Analytics Software and Solutions,* www.sas.com/en_us/home.html

[17] *IBM SPSS Statistics,* www.ibm.com/products/spss-statistics

[18] *Tableau,* www.tableau.com/

[19] *The Great Influenza,* www.amazon.com/Great-Influenza-Deadliest-Pandemic-History-dp-0143036491/dp/0143036491/

CHAPTER 5

Good Datasets

Publishing data in a reusable form to support findings must be mandatory.

—Royal Society [1]

In the previous chapters, we learned about *data items* (analysis variables) – the components and contents of datasets. But a major decision for your research is *how* that data is to be formatted and stored and especially how to make it easy to analyze and share. Poorly designed and implemented datasets are just as problematic as any bad data they might contain. It's hard to get good data out of bad datasets, and that can make life difficult for other analysts who want to use your data.

Next, we learn about the various ways to format and store data, reviewing several commonly used formats. Most of the programming and analysis tools discussed earlier have helpful methods and add-on libraries for creating, reading, analyzing, and converting dataset formats. We show some examples of these processes using Python and other tools.

Sharing Data

Research data and results should be shared. This allows others to evaluate your claims, encourages collaboration, and can enable those who are smarter than you to advance your research.

Data analytics is a scientific endeavor. That means it should follow the traditions of scientific analysis, which means encouraging your methods and results to be replicated, confirmed, and even extended. Therefore, you generally need to share your data in some manner, that is, documenting and structuring your dataset so that other researchers can find it, understand it, reuse it, and even combine it with data from other

© Harry J. Foxwell 2020
H. J. Foxwell, *Creating Good Data*, https://doi.org/10.1007/978-1-4842-6103-3_5

studies. This obviously encourages you to *create good data* from the start, avoiding the embarrassment of publishing bad data.

Good reasons for *not* sharing your data include proprietary or classified research, privacy concerns and regulations, or the need to establish scientific priority. But even these types of data will need to be of sufficient quality that it can be shared with authorized users.

Note   Important scientific and social research data can "disappear" – become inaccessible – sometimes for political reasons. Datasets on climate change and on political unrest have been removed from public availability. Independent archiving and sharing of inconvenient or controversial data and research is often necessary so that valuable work is not lost.

Several organizations support the concept of FAIR data [2]. Such data needs to be

- **F**indable: Uniquely identified, completely described, and publicly registered or indexed in a searchable resource

- **A**ccessible: Retrievable using standard protocols

- **I**nteroperable: Using broadly applicable representations

- **R**eusable: With clear usage rules and accurate metadata

Following these guidelines adds value and credibility to your research and connects you to collaborative communities relevant to your domain of interest. NIST goes so far to claim, "If data providers published FAIR data that are *analysis ready*, data users would not need to spend 70-80% of their time on data preparation." [3]

Dataset Dictionaries/Metadata

Every dataset needs a *dictionary* that at minimum describes its contents and additionally includes references to its owner/creator and its location.

> *Metadata should be written for a user twenty years into the future who is unfamiliar with the project, sites, methods, or observations.*

> —*J. Michaud and J. Friddell [4]*

While it's generally a best practice to prepare and provide metadata, depending on the nature of your research support, you might even be *required* by policy or by law to prepare and provide such publicly available metadata and other documents.

Good Metadata

Good metadata helps you, your readers, and other analysts to thoroughly understand the source of your investigation and conclusions. It theoretically should allow other researchers to duplicate your methods and results, so *each dataset's metadata* must therefore include as much detail and description as possible, including at least

Data items (a.k.a. "variables")

- Item names and narrative descriptions

- Data types (both computational and analytical)

- Units of measurement

- Allowable range of values

- Special encodings (e.g., Unicode)

Dataset file

- Creator/owner

- Creation date

- File type and size

- Usage/privacy/sharing restrictions

- Reference/location

    - See `https://infoguides.gmu.edu/citingdata` [5].

There are efforts to standardize data documentation, and as you might expect, different knowledge disciplines each having their own vocabularies and recommendations for creating metadata. For example, the *Dublin Core Metadata Initiative* [6] defines general standards and formats for describing data of all types and sources, ands the *Digital Curation Centre* [7] proposes best practices for managing datasets (their tagline is *because good research needs good data*), as well as suggesting domain-specific metadata standards for fields such as Physical, Biological, and Earth Sciences, Social Sciences, and Humanities.

What's in a Name?

Start with *good names* for your data items and datasets.

Most professional data analysts use a variety of tools in their work – different programming languages like R, Python, and SQL – hosted on different operating systems like Windows, OS X, and Linux. Ideally, datasets and data items should use item and file naming conventions which are *interoperable* across all combinations of those environments. That is, although Windows and OS X accept file names containing spaces and non-letter characters (i.e., "`This is an evil file name!.txt`"), Linux can have difficulty dealing with such obvious abominations when parsing with shell scripts and other programs. The same can be true of data item names, which can also cause parsing problems for programming tools along with lengthy and complex dataset debugging puzzles.

Data Item Naming

Good dataset format starts with something very basic: what to call your data items. Data names should be unique, relatively short, mnemonic (easily calling to mind the referent of the measurement), and consistent with the requirements of the tools used to analyze their values. Obscure or cryptic item names hinder understanding of the data and can lead to delays and errors in analysis.

For example, in naming an item that represents the *weight* of an individual, you might consider

1. `W`

2. `VAR15`

3. `Patient's Weight`

4. `pweight`

5. `PATWT`

6. `Mxyzptlk` [8]

While one or more of these item/variable names might actually work in an analytical language's syntax, some names, like #1, could be confusing or ambiguous. Numbers #2, #4, #5, and #6 are likely legal names, but #2 is not mnemonic of the actual measurement and would need frequent lookup or reminding of its measurement meaning, especially if there were a large number of data items. And never use spaces and non-alphabetic characters in item naming, like #3, even if they are allowed by the primary analysis tool; such syntax will eventually cause problems when accessed by other tools. Avoid cute or esoteric names like #6 which might be humorous or entertaining at first but have no place in serious research.

Even with detailed documentation in your metadata (data dictionary), there are still analytical issues that depend on the specific tools used to ingest the data that are not addressed in the metadata. Such dependencies include, for example, whether the tool is *case sensitive* with respect to data names and its restrictions of special characters or naming formats. Python data names are case sensitive ("Weight" ≠ "weight"), while those in SQL are often not (depending on the database implementation!). Most analytical tools rightly prohibit *evil* characters such as spaces or slashes in data names. Analysts can spend much time debugging data ingest issues caused by odd data naming choices. A good rule, then, is keep such names clear and simple; don't get too fancy or creative. And ensure that you or your data analyst thoroughly understands the naming requirements of the programming tools used on the dataset.

Dataset File Naming

Like data items, dataset names should be unique, short, and descriptive, containing no bothersome (non-letter/non-numeric) characters. One goal for such naming is *interoperability* among any OS environments you or your users might run. To reiterate, file names with embedded blanks, for example, can be problematic for UNIX-like operating systems.

Dataset Formats

Even before you start collecting data, you need to consider how you are going to package it to enable easy input into your analytical tools, as well as making it convenient for others to use. In this section, we review a few of the many ways to organize data into file formats that enhance "analyzability" and "shareability."

Keep It Simple

There are quite literally dozens of "standard" formats for collecting and storing research and other data [9]. And there is no *one-size-fits-all* format that can fully represent the variety of data types and structures found in complex research data. Nevertheless, there are several common formats which enable simple and clear presentation of most data collections.

Comma-Separated Values (.csv)

The most frequently encountered dataset format is a *plain text file* consisting of a *header line* listing the collected data item names, followed by the individual data records [10]. The names and data items are "separated" by a *delimiter* character, usually a *comma* (",") by default. However, such files can often use alternate delimiters like semicolons (";"), vertical bars ("|"), or tab characters, especially if the data itself contains commas. Figure 5-1 shows the first few lines of a typical .csv file.

```
recno,abv,ibu,id,name,style,brewery_id,ounces
0,0.05,,1436,Pub Beer,American Pale Lager,408,12.0
1,0.066,,2265,Devil's Cup,American Pale Ale (APA),177,12.0
2,0.071,,2264,Rise of the Phoenix,American IPA,177,12.0
3,0.09,,2263,Sinister,American Double / Imperial IPA,177,12.0
4,0.075,,2262,Sex and Candy,American IPA,177,12.0
5,0.077,,2261,Black Exodus,Oatmeal Stout,177,12.0
6,0.045,,2260,Lake Street Express,American Pale Ale (APA),177,12.0
7,0.065,,2259,Foreman,American Porter,177,12.0
8,0.055,,2258,Jade,American Pale Ale (APA),177,12.0
```

Figure 5-1. `beers.csv` *data file of craft beers [11]*

Note the use of short, mnemonic data item names in the header line, the data records that follow, and, of course, *commas* as the delimiters. Each record must contain the same number of data values as the number of named items in the header; missing data is often indicated using a character string such as "NA" (Not Available) or an empty field bracketed by the delimiter (",,"). Data values must not contain the assigned delimiter character, and there should be no "trailing delimiter" at the end of any record lines. Such inclusions will cause errors when the dataset is imported into programming tools like R and Python and might generate confusing error messages that take extra time to understand and resolve.

Properly formatted datasets are easy to import into R or Python. R has a built-in function for reading .csv files:

```
> BeerData <- read.csv("beers.csv", sep=",")
> head(BeerData)
```

recno	abv	ibu	id	name	style	brewery_id	ounces
1	0 0.050	NA	1436	Pub Beer	American Pale Lager	408	12
2	1 0.066	NA	2265	Devil's Cup	American Pale Ale (APA)	177	12
3	2 0.071	NA	2264	Rise of the Phoenix	American IPA	177	12
4	3 0.090	NA	2263	Sinister	American Double/Imperial IPA	177	12
5	4 0.075	NA	2262	Sex and Candy	American IPA	177	12
6	5 0.077	NA	2261	Black Exodus	Oatmeal Stout	177	12

Python uses the separately loaded *pandas* library function to do the same, creating a *data frame*:

```
>>> import pandas as pd
>>> beerdata = pd.read_csv('beers.csv', sep=',')
>>> beerdata.head()
```

	recno	abv	ibu	...	style	brewery_id	ounces
0	0	0.050	NaN	...	American Pale Lager	408	12.0
1	1	0.066	NaN	...	American Pale Ale (APA)	177	12.0
2	2	0.071	NaN	...	American IPA	177	12.0
3	3	0.090	NaN	...	American Double / Imperial IPA	177	12.0
4	4	0.075	NaN	...	American IPA	177	12.0

Note   In this book, we do not cover the analysis phase of data exploration in depth, only the data preparation practices and tools. For advice on analytics, see the Apress publications at www.apress.com/us/search?query=data+analytics.

Unfortunately, perhaps, data doesn't always come in simple .csv files. Microsoft Excel is often used to record and save data, but both R and Python have libraries and methods to read data directly from spreadsheets. Excel can also *export* its data to .csv files (see its *Save As* menu item). However, you'll need to review the resulting output for proper header and item naming.

CSV files are an example of "structured data" – having named fields organized into records – and are the simplest form of such datasets. But often data item relationships and repeated measurements within categories result in more complex hierarchical formats.

That being said, avoid embedding compound data items in your records. That is, each named data item should encode *only one value* for that item. A good example of this problem is the well-known IMDb dataset [12]. The Genre data item lists multiple values for each movie record. Extracting counts for each genre requires extra parsing of that item; more work for the data analyst!

JSON Datasets

The .csv format is a "flat" file structure, one complete record per line, with limited ability to represent hierarchically organized content. JSON (JavaScript Object Notation) is another quite common dataset format, similar to XML [13]. It is generally used for *nested* data representations, having multilevel records with varying numbers of subrecords. For example, see the public domain football (soccer) datasets at https://github.com/openfootball/football.json and a sample at https://raw.githubusercontent.com/openfootball/football.json/master/2015-16/en.1.json (not reprinted here due to space limitations).

Python has a json library for importing this dataset format [14]. All the previously discussed rules for naming data items and datasets still apply!

HTML Data

While few researchers create and store data directly into HTML [15] pages, the Web is full of interesting and useful datasets presented in that format. See, for example, www.top500.org/lists/2019/11/, the list of the world's most powerful supercomputers. The list's authors maintain a database of the system characteristics, and the data is rendered from that source into an HTML table. Such tables can be converted to simpler .csv files using various "screen scraping" libraries, notably *Beautiful Soup* [16], which parses the HTML <table> markup and identifies each of the embedded <tr> and <td> table data item values.

More Dataset Formats

A quick review of major dataset providers such as `www.data.gov/` reveals myriad other dataset distribution formats like XML, PDF, Text, ESRI, ODF, R, SAS, SPSS, and numerous database formats. Your choice of how to capture and represent your data will vary according to its structure, types, and the software tools used to record it. In all cases, follow good metadata practices for naming and documenting your data. And look for software tools and libraries, particularly in the popular Python language, which can easily import those formats.

Remember: *Keep it simple*!

Managing Datasets

You need to ensure the quality of the data within a dataset and the appropriate formatting of the file and metadata. But wait, there's more! Datasets are *valuable assets* – "the new oil" according to UK mathematician Clive Humby [17]; and like oil must be refined and cared for or "curated" to ensure their usefulness, that is, classifying their *value* to the researchers or organizations who use them.

There are several key ways to classify datasets with respect to their *value* and *usage*:

- Retention policy

  - Is your data temporary for an ad hoc project, or does it need to be kept for some defined period (or forever)?

- Confidentiality policy

  - Who is permitted to see, use, copy, share, or change the data? Is the data formally identified as internal use, trade secret, classified?

- Level of confidence

  - Is the data raw and unprocessed, verified, timely, non-obsolete?

- Value

  - How critical is the data to the business or researcher? What if it is lost; no problem, or catastrophe?

In cases where data is collected on individual people, you must ensure that data collected for one purpose cannot be used for any other purpose without the consent of the data subjects, and that *reidentification* of individuals in supposedly anonymized data is limited [18]. Such restrictions are typical for most academic and government-funded human-subject research, even for basic surveys.

All of these questions must be answered and addressed; if the data has value, it must be protected, through encryption, archiving, duplication, and monitoring. And don't forget that storage media and software formats decay and become obsolete; datasets on floppy disks (if there are any readers old enough to remember those!) are essentially extinct.

E pluribus unum? [19]

If you are collecting data in multiple sessions or categories, you might be tempted to create separate dataset files for each instance. If the data items collected for each instance are exactly the same, there is really no reason to use multiple files; just add a data item indicating the instance. Admittedly, combining datasets with identical record structures is trivial, for example, using Python's pandas library concat() function [20] which basically just glues multiple identically formatted files together, but if there are record structure differences among the dataset instances, combining them requires extra time and effort as well as using (or creating) a *key* data item and then using the pandas merge() function.

If you do choose to present multiple related datasets, it is helpful to combine them along with metadata and other files into a compressed archive. Standard tools for such aggregation include zip and tar, which are generally compatible with various operating systems; *compression* should be used for very large files.

Is Your Data *Ready*?

Your ultimate goal in creating data for analysis for yourself or for others is to provide a resource that is *ready* for research purposes. Neil Lawrence of the University of Cambridge proposes three major levels of "data readiness," primarily for Data Science research and for business analytics, but relevant nevertheless to general dataset preparation for any kind of project [21]. He says data that is *ready* must be *available* in a commonly *usable* format, be *loadable* into analysis software, and, more importantly, is *appropriate* to answer the questions for which it was gathered. Good advice!

Summary

We have covered quite a bit in this chapter, focusing on *creating good datasets*. We emphasized the importance of this not only for your own work but for the inevitable *sharing* of your data. Sharing requires *good metadata*, usable dataset formats, interoperable naming conventions, and effective curation practices that ensure the readiness of your data for analysis.

Next, we address a major impediment to creating good data – *bias* in data collection.

Chapter References

[1] *Final report - Science as an open enterprise*, https://royalsociety.org/topics-policy/projects/science-public-enterprise/report/

[2] *"A love letter to your future self": What scientists need to know about FAIR data*, www.natureindex.com/news-blog/what-scientists-need-to-know-about-fair-data

[3] *NIST Big Data Interoperability Framework: Volume 9, Adoption and Modernization Version 3*, https://nvlpubs.nist.gov/nistpubs/SpecialPublications/NIST.SP.1500-10r1.pdf

[4] *Best Practices for Sharing and Archiving Datasets*, www.polardata.ca/pdcinput/public/PDC_Best_Practices_FULL.pdf

[5] *Citing Datasets*, https://infoguides.gmu.edu/citingdata

[6] *Dublin Core Metadata Initiative*, www.dublincore.org/about/

[7] *Digital Curation Centre*, www.dcc.ac.uk/resources/metadata-standards

[8] *Mister Mxyzptlk*, https://superman.fandom.com/wiki/Mister_Mxyzptlk

[9] *Format Descriptions*, www.loc.gov/preservation/digital/formats/fdd/browse_list.shtml

[10] *CSV, Comma Separated Values (RFC 4180)*, www.loc.gov/preservation/digital/formats/fdd/fdd000323.shtml

[11] *Craft Beer Dataset,* www.kaggle.com/nickhould/craft-cans

[12] *IMDb Datasets,* www.imdb.com/interfaces/

[13] *JSON (JavaScript Object Notation),* www.loc.gov/preservation/digital/formats/fdd/fdd000381.shtml

[14] *JSON encoder and decoder,* https://docs.python.org/3/library/json.html

[15] *HyperText Markup Language (HTML) Family,* www.loc.gov/preservation/digital/formats/fdd/fdd000475.shtml

[16] *Beautiful Soup Documentation,* www.crummy.com/software/BeautifulSoup/bs4/doc/

[17] *Clive Humby,* https://towardsdatascience.com/data-is-not-the-new-oil-bdb31f61bc2d

[18] *Sharing Data: What to Do with Your Processed Data,* https://responsibledata.io/resources/handbook/chapters/chapter-02c-sharing-data.html

[19] *E pluribus unum,* www.dictionary.com/browse/e-pluribus-unum

[20] *Merge, join, and concatenate,* https://pandas.pydata.org/pandas-docs/stable/user_guide/merging.html

[21] *Data Readiness Levels,* http://data-readiness.org/

Good Data Collection

Our preferences do not determine what's true.

—Carl Sagan [1]

We have learned how to create good data items and dataset formats, focusing primarily on the characteristics of the individual data items and on the organization of the datasets. But there is a big issue not yet discussed – *how to choose* what data items to include in your research and *how to collect* appropriate values for those items. This process is fraught with danger – the danger of *bias*. Bias is the data collection killer; nothing will compromise the quality of your data more than bias. Biased data leads to incorrect results and faulty conclusions.

Good data minimizes bias.

What Is *Bias*?

With respect to data gathering and collection, bias is any intentional or unintentional *preference* toward an analytical process that can lead to false or misleading conclusions. *Therefore, bias is an error in your research.* Your data should speak the truth, not your wishes.

The goal of research is to conduct measurements and observations of some occurrence or phenomenon with the intent of determining its *true behavior and characteristics*. With few exceptions, we cannot observe all possible outcomes nor all members of a population under study. At best, we gather a *subset* or *sample* of measurements and then attempt to extrapolate from analytical results to the entire universe or population. That is where bias sneaks in.

Note   The most important step in minimizing bias is to honestly *admit* that it is possible and even *quite likely* to exist in your research data and analysis and to thoroughly search for and eliminate its sources.

It is easy to discover examples of data collection bias in popular media reports. Surveys of voter sentiment and opinion studies in social media are particularly susceptible to this effect due to faulty selection methods. For example, calling listed landline phone numbers for political campaigns misses important population segments – those with only cell phones (an increasing trend) – or mass email queries about support for various activist issues, missing those without such technology or those who choose to ignore such communications. These are examples of *selection bias* – omitting key components of the intended target populations either through error, ignorance, or even deceit and resulting in overrepresentation or underrepresentation of important population subsets. The archetypal example of this, and most often mentioned in statistics textbooks, is the 1948 Chicago Tribune presidential election poll – Dewey vs. Truman [2], where the Tribune announced the wrong winner based on a biased poll.

Major Types of Bias

Beliefs and preferences affect decisions about what to measure, who to ask, and how to record. Researchers might choose methods, measurements, and tools that conform to their cognitive needs and backgrounds. Researchers are often unaware of those effects on their choices.

The inability to perceive something, or to perceive incorrectly, is called *cognitive bias*. It affects many areas of life, including prejudicial thoughts and behaviors, social and association choices, and, yes, research projects. In part it arises from the very nature of how our brains function. Our visual and reasoning senses are attuned to what is essential to our survival, filtering out "irrelevant" data. Wikipedia lists more than 100 subtle variations of cognitive bias [3], some of which apply to data collection and analysis such as *expectation bias*, where results that conform to how your data *should* appear are accepted and those that don't are rejected or ignored.

One of the most common cognitive biases is *confirmation bias* – selecting data definitions and collection procedures that conform to a preconceived view of the research area under study. In research, we often come up with an idea to investigate, and even there bias starts since we are already making some conclusions about what to measure and what is irrelevant. For example, when considering investigating the factors which affect blood pressure, we might assume that diet and weight are involved and that eye color and shoe size are not. Properly, even such assumptions should be examined and tested; you might be surprised!

Sampling Bias

We sample because we generally can't explore an entire population. But to get *good data*, we need a *good sample*.

Researchers must try to thoroughly understand the populations or universes from which their data or cases are being drawn. And the collected subset, or *sample*, ideally should be *representative* of the target population. That means the breakdown of the sample by various categories should reflect as much as possible the *same distribution* of those categories in the population (assuming they are known). For example, a *random* sample drawn from the US population, which is approximately 51% female, should itself be about 51% female. There's that special word *random*. How can you achieve truly random sampling and avoid sampling bias? By ensuring that your collection or measurement protocols do not overlook or overselect important data categories. And a researcher's preferential selection might cause such variance from randomness, as when a male researcher intentionally or unconsciously selects subsets predominantly male.

What Does "Random" Selection Mean?

Good sampling requires selecting subsets from a population where *each item has the same probability of being selected*. When this doesn't happen, there is *bias*. We won't go into the statistical theory of randomness here, other than to say that true randomness is difficult to achieve, but we must try anyway. We can generally trust functions such as the Python *random* module [10] or the pandas library's `sample()` function [11] to provide reasonably equal probability selection. Although some computer functions that purport to be random are known to not be ideally random.

Good Sampling

What is *good sampling* (or good data collection)? A good sample has the quality of *representativeness* – having as closely as possible the same characteristics as the population from which it was selected. Then, descriptions and explanations derived from an analysis of the sample may be assumed to represent similar ones in the population. And larger samples, by proportion, are better than smaller ones which yield greater uncertainty of results. Your sampling and measurement methods will be determined by your analytical goals, such as estimating the statistical characteristics and data item interrelationships of the population.

Note   Recording data using instruments is like taking a sample from the population of all possible measurements. So, treat such data as a sample, and follow good sampling rules.

More Data Collection Problems

While confirmation and sampling biases are among the most common data collection issues, there are myriad other potential dangers that can bias your data. Collecting data over long periods of time can result in *temporal bias* effects and errors, where data gathered at the start of the study differs significantly from data gathered later, due to boredom, learning effects, or changes in understanding the purpose of the study, not to mention that there might be true temporal effects in the data. There is also a temptation to keep measuring or collecting until some impression that "enough" has been gathered rather than prior planning for determining sample size. Worse still are non-technical pressures for gathering sufficient or relevant data due to research deadlines, peer pressure, or policy constraints.

Here are just a few more obstacles to creating unbiased research data:

- Funding/conflict of interest
  - Who is paying for your research?
  - Do they have influence on your resources, methods, conclusions, or publications?

- Small samples

  - Reliable statistical hypothesis tests require sufficient sample sizes.

- Availability/convenience bias

  - Choosing selection methods or measurements because they are easy to obtain rather than specific and relevant.

    - Consider how much research is conducted using volunteer/paid college students!

  - Obtaining measurements/answers only from those willing to respond or participate.

- Expectancy bias

  - Interpreting or adjusting measurements influenced by prior beliefs.

- Measurement bias

  - Using faulty instruments to obtain data.

    - Test and validate instruments before measuring!

Generally, there are two forms of sampling: probabilistic and non-probabilistic. The first form depends upon random selection methods, aiming to result in representativeness of the population from which the sample is drawn; this is not always easy and requires careful procedures and sample sizes. The second form uses deliberate (non-random) choices as to what population members are selected (usually to guarantee representativeness); this is often better than probabilistic sampling but requires better understanding (and availability) of information about the characteristic of the population.

Quota sampling: Random sampling within each characteristic of interest of the population under study so that the sample's distribution of those characteristics matches that of the population. For example, if you know that your target population is 20% unemployed and 80% employed, identify and select those same proportions for your sample. Otherwise, simple random sampling from the population might accidentally result in underrepresentation of one of those groups.

Homogeneous sampling: Identifying a population with known characteristics and sampling from a subset, all of which have the same characteristic distribution, for example, pre-identifying female college graduates and then random sampling from that population.

Casual/ad hoc sampling: No criterion for choosing cases, just what happens to attract your attention in the target population. Probably the worst method, as it is likely most affected by the biases we discussed earlier.

All this means that indeed there are many possible ways for you to fail at obtaining unbiased data. So you need to find them and address them.

Recognizing and Reducing Bias

Notice that we said *reducing* bias, not *eliminating* it. It's difficult, if not impossible, to be *perfectly* fair and objective due to unconscious biases that you are not even aware of. *Awareness*! That bias exists, and that you can suffer from it, even unintentionally.

Recognizing bias in research comes from training, experience, and depth of knowledge of the domain of study. *Peer review* or expert oversight is essential during research design and execution; groups are generally better at finding errors and bias problems than are individual researchers. And taking a contrary view of your own research, being open to disproving your hypothesis, is also a good and honest way to reveal and address certain forms of bias.

While we don't cover the analytical phase of research in detail in this book, we'll mention a couple typical bias-producing behaviors or actions that can arise in the collection phase as the data is being examined.

Understanding Outliers

Not all outliers are errors. Outliers are often thought of as extreme or nonsensical recorded measurements that exceed the expected range of possible values. For example, in a survey of new high school graduates, many of the reported ages were under 20 years, but one of the reported responses was 91 years. There's a temptation to dismiss this as a transposition typo (should be 19?), until you read about people finally graduating after many decades without their high school diploma [4].

Outlier analysis is a basis for monitoring business transactions like credit card purchases. A customer who lives in Minnesota (MN) shows a transaction in New Mexico (NM). Is it a transposition error, fraud, or a vacation purchase? You should thoroughly examine all outliers to determine if they are indeed errors or if they reveal something surprising about your research subject. Isaac Asimov's observation [5] that "the most exciting phrase to hear in science, the one that heralds new discoveries, is not 'Eureka!' but 'That's funny…'" applies to outliers! Disbelief and dismissal of outliers is a common form of bias.

The Consequences of Bias

Biased results and conclusions come from bad (biased) data. And if your research conclusions are used for determining policies or actions that affect people's lives, you are *obligated* to base them on *good data*. Bad data has resulted in discriminatory hiring and recruiting decisions at Amazon [6] based on machine learning algorithms fed with biased training data. Banks' algorithms have rejected loan applications of minorities due to the same type of problem [7]. Biased historical data used for AI sentencing algorithms has denied bail and recommended longer prison stays [8]. And poorly trained algorithms have resulted in self-driving car fatalities [9].

Bad data can have bad consequences.

Summary

In this chapter, we alerted you to the many ways your data collection efforts can be infected with various types of bias. Your awareness of the likelihood and pervasiveness of bias in research is necessary to encourage actions that will result in good data.

Next, we'll examine a few example datasets that illustrate both good and bad data practices.

Chapter References

[1] *Carl Sagan*, https://simple.wikiquote.org/wiki/Carl_Sagan

[2] *Dewey Defeats Truman: How Sampling Bias can Ruin Your Model*, https://medium.com/@ODSC/dewey-defeats-truman-how-sampling-bias-can-ruin-your-model-f4f67989709e

[3] *List of Cognitive Biases*, https://en.wikipedia.org/wiki/List_of_cognitive_biases

[4] *Oldest High School Graduates in the World*, www.oldest.org/culture/high-school-graduates/

[5] *Isaac Asimov Funny Quotes*, www.azquotes.com/quote/11682

[6] *Amazon ditched AI recruiting tool that favored men for technical jobs*, www.theguardian.com/technology/2018/oct/10/amazon-hiring-ai-gender-bias-recruiting-engine

[7] *Credit denial in the age of AI*, www.brookings.edu/research/credit-denial-in-the-age-of-ai/

[8] *Courts Are Using AI to Sentence Criminals. That Must Stop Now*, www.wired.com/2017/04/courts-using-ai-sentence-criminals-must-stop-now/

[9] *What Happens When Self-Driving Cars Kill People?*, www.forbes.com/sites/cognitiveworld/2019/09/26/what-happens-with-self-driving-cars-kill-people/#18657013405c

[10] *Python Random Module*, https://pynative.com/python-random-module/

[11] *pandas.DataFrame.sample( )*, https://pandas.pydata.org/pandas-docs/stable/reference/api/pandas.DataFrame.sample.html

Dataset Examples and Use Cases

A precise, well-formed definition is one of the most critical requirements for shared understanding of data [1]

—ISO/IEC 11179-1

To illustrate some of the practices we learned about in the preceding chapters, let's have a brief look at the characteristics of a few example datasets. We will examine some well-known datasets that pop up regularly in data analytics courses and textbooks for teaching about visualization, regression modeling, machine learning, and other data exploration methods. There are literally thousands of online datasets hosted by independent researchers, government agencies, corporations, and academics. These online datasets vary widely in the completeness and quality of their metadata and of the data itself and are worth reviewing for their ranges of "goodness."

The Titanic Survivor Dataset

On Sunday, April 14, 1912, the passenger ocean liner RMS Titanic, the largest such ship at that time, struck an iceberg in the North Atlantic and sank in less than 3 hours. More than 1500 of her 2224 passengers and crew perished [2]. The dataset about the survivors is perhaps one of the most cited and studied in data analytics courses [3,4,5] and is used to illustrate machine learning algorithms, cluster analysis, and basic statistical and visualization methods using R and Python. We highlight it here because of its excellent *metadata documentation* which includes its origin, content, detailed variable descriptions, and discussion of missing values. For example, the data item names and descriptions are shown here:

```
VARIABLE DESCRIPTIONS:
pclass          Passenger Class
                (1 = 1st; 2 = 2nd; 3 = 3rd)
survival        Survival
                (0 = No; 1 = Yes)
name            Name
sex             Sex
age             Age
sibsp           Number of Siblings/Spouses Aboard
parch           Number of Parents/Children Aboard
ticket          Ticket Number
fare            Passenger Fare
cabin           Cabin
embarked        Port of Embarkation
                (C = Cherbourg; Q = Queenstown; S = Southampton)
boat            Lifeboat
body            Body Identification Number
home.dest       Home/Destination
```

Note the simple, descriptive data item names, followed in the metadata file by fuller definitions of their meanings, and a well-formed CSV dataset ready to be loaded into an R or Python data frame.

The IBM Employee Attrition Dataset

Another well-known and frequently studied *synthetic* dataset (fictional, created for study purposes), this data was created to help model machine learning and predictive analytics [6]. Presented in CSV format, it includes 35 well-named (i.e., self-defining) data items, 1470 records, and no missing values. Data items include all our previously studied analytical types (nominal, ordinal, interval, and ratio), including purposeful outliers (for practice!). Notice the use of clearly descriptive variable names in the header with no non-letter characters:

```
Age,Attrition,BusinessTravel,DailyRate,Department,DistanceFromHome,Educatio
n,EducationField,EmployeeCount,EmployeeNumber,EnvironmentSatisfaction,Gend
er,HourlyRate,JobInvolvement,JobLevel,JobRole,JobSatisfaction,MaritalStatu
s,MonthlyIncome,MonthlyRate,NumCompaniesWorked,Over18,OverTime,PercentSala
ryHike,PerformanceRating,RelationshipSatisfaction,StandardHours,StockOpti
onLevel,TotalWorkingYears,TrainingTimesLastYear,WorkLifeBalance,YearsAtCo
mpany,YearsInCurrentRole,YearsSinceLastPromotion,YearsWithCurrManager
```

This Kaggle dataset includes additional information and some summary analytics about the dataset, including links to discussions about the data and how to explore it. Kaggle is a popular online community of data analysts and contributors who freely contribute datasets and example analyses and who sponsor competitions of challenge problems especially those involving machine learning and artificial intelligence methods [11].

The Internet Movie Database (IMDb)

The Internet Movie Database is a very popular source for information and reviews about movies, TV shows, and other streaming media content. It has millions of viewer contributed records about such content, including searchable datasets of descriptions and ratings [7]. The datasets are not only favorites of movie viewers but also of data analysts and course instructors. The dataset metadata is well-documented:

title.basics.tsv.gz - Contains the following information for titles:
      tconst (string) - alphanumeric unique identifier of the title
      titleType (string) – the type/format of the title (e.g. movie,
      short, tvseries, tvepisode, video, etc)
      primaryTitle (string) – the more popular title / the title used by
      the filmmakers on promotional materials at the point of release
      originalTitle (string) - original title, in the original language
      isAdult (boolean) - 0: non-adult title; 1: adult title
      startYear (YYYY) – represents the release year of a title. In the
      case of TV Series, it is the series start year
      endYear (YYYY) – TV Series end year. '\N' for all other title types

runtimeMinutes – primary runtime of the title, in minutes

genres (string array) – **includes up to three genres associated with the title**

As good as these datasets are, there are a few items that require extra cleaning or transformation before analysis. We recommended against compound data items in Chapter 2; the Genre and Actors values are just such cases:

Rank,Title,**Genre**,Description,Director,**Actors**,Year,RuntimeMinutes,Rating, Votes,RevenueMillions,Metascore

1,Guardians of the Galaxy,**Action;Adventure;Sci-Fi**,A group of intergalactic criminals are forced to work together to stop a fanatical warrior from taking control of the universe.,**James Gunn,Chris Pratt; Vin Diesel; Bradley Cooper; Zoe Saldana**,2014,121,8.1,757074,333.13,76

2,Prometheus,**Adventure;Mystery;Sci-Fi**,Following clues to the origin of mankind; a team finds a structure on a distant moon; but they soon realize they are not alone.,Ridley Scott,**Noomi Rapace; Logan Marshall-Green; Michael Fassbender; Charlize Theron**,2012,124,7,485820,126.46,65

While this might seem easily fixed (transformed into new variables with separate indicators of genre and actors), this is yet another cleanup task imposed upon the analyst's helper's time and programming skills.

US Hurricane Data

Those who record data don't always think about future analysts who want to explore that data. Here's a case of important weather/climate data on hurricane activity presented in an HTML table [8]. It's easy enough to scrape the data into a CSV file (e.g., using the BeautifulSoup library), but, again, extra work for the data analyst due to long variable names with embedded spaces and compound data representations for States_Affected. The metadata is very well done (in the site's "Notes" section), but it's a challenge parsing that column, requiring new variable creation:

Year|Month|**States_Affected**|Highest_Category|Central_Pressure_mb|Max_Winds_kt|Name

1851|Jun|**TX,C1**|1|977|80|NA

1851|Aug|**FL,NW3;I-GA,1**|3|960|100|"Great Middle Florida"

1852|Aug|**AL,3;MS,3;LA,2;FL,SW2,NW1**|3|961|100|"Great Mobile"

```
1852|Sep|FL,SW1|1|985|70|NA
1852|Oct|FL,NW2;I-GA,1|2|969|90|"Middle Florida"
1853|Oct|GA,1|1|965|70|NA
1854|Jun|TX,S1|1|985|70|NA
1854|Sep|GA,3;SC,2;FL,NE1|3|950|100|"Great Carolina"
```

UFO Sighting Data

And now for an example that's out of this world. There is a growing database of nearly 80,000 UFO sightings dating back more than 60 years [9,10]. Who knew? And the sightings are meticulously documented (good metadata!). The dataset contains fascinating details about our supposed ET visitors such as object shapes (saucer, disk, cylinder, spheres, cubes, cigars, etc.). But the dataset could make things a bit easier for analysts by using simpler variable names (without special characters) and by recording duration in consistent units (although the reported time has been converted into an extra seconds variable):

datetime,city,state,country,shape,duration (seconds),**duration (hours/ min),**comments,date posted,latitude,longitude

10/10/1949 20:30,san marcos,tx,us,cylinder,2700,**45 minutes**,"This event took place in early fall around 1949-50. It occurred after a Boy Scout meeting in the Baptist Church. The Baptist Church sit",4/27/2004,29.8830556,-97.9411111

10/10/1949 21:00,lackland afb,tx,,light,7200,**1-2 hrs**,"1949 Lackland AFB&#44 TX. Lights racing across the sky & making 90 degree turns on a dime.",12/16/2005,29.38421,-98.581082

10/10/1955 17:00,chester (uk/england),,gb,circle,20,**20 seconds**,"Green/ Orange circular disc over Chester&#44 England",1/21/2008,53.2,-2.916667

10/10/1956 21:00,edna,tx,us,circle,20,**1/2 hour**,"My older brother and twin sister were leaving the only Edna theater at about 9 PM&#44...we had our bikes and I took a different route home",1/17/2004,28.9783333,-96.6458333

10/10/1960 20:00,kaneohe,hi,us,light,900,**15 minutes**,"AS a Marine 1st Lt. flying an FJ4B fighter/attack aircraft on a solo night exercise&#44 I was at 50&#44000&#39 in a "clean" aircraft (no ordinan",1/22/2004,21.4180556,-157.8036111

Lessons Learned

We see from the preceding examples the methods that researchers and other data collectors use to record their observations in ways that can either help or hinder analysis. Good metadata, using good naming conventions, thinking ahead about how the data should be recorded, and avoiding complex data representations, will save much of analyst's time and effort.

Useful Dataset Sources

In addition to the datasets mentioned earlier, there are numerous sources of research data which can serve as inspiration and examples of both good and bad data collection and metadata practices. They are worth reviewing for their general interest as well as sources for demonstrating specific analytic methods like machine learning, visualization, and model building:

- The home of the U.S. Government's open data, `www.data.gov/`

- Explore US Census Data, `www.census.gov/data.html`

- World Bank Open Data, `http://data.worldbank.org/`

- Registry of Open Data on AWS, `https://aws.amazon.com/public-datasets/`

- Google Cloud Public Datasets, `https://cloud.google.com/public-datasets/`

- Welcome to Wikidata, `www.wikidata.org/wiki/Wikidata:Main_Page`

- Applied Regression Analysis and Generalized Linear Models, Third Edition: Data Sets, `https://socialsciences.mcmaster.ca/jfox/Books/Applied-Regression-3E/datasets/`

- UCI Machine Learning Repository, `https://archive.ics.uci.edu/ml/datasets.php`

Summary

In this chapter, we have presented some typical examples of popular and research datasets. We can use the variety and quality of their data and metadata as examples to emulate or to avoid. We focused on inefficient variable naming and use of compound data items as particularly time-consuming to transform; prior planning and good data practices will help you minimize these problems.

In our next chapter, we admit that not everyone follows best practices for data representation and collection. And so, we review several key methods for cleaning up such datasets.

Chapter References

[1] *Information technology — Metadata registries (MDR) — Part 1: Framework*, https://standards.iso.org/ittf/Publicly AvailableStandards/c061932_ISO_IEC_11179-1_2015.zip

[2] *Sinking of the RMS Titanic*, www.britannica.com/topic/Titanic

[3] *Data obtained from Vanderbilt University Department of Biostatistics*, http://biostat.mc.vanderbilt.edu/wiki/Main/DataSets

[4] *Titanic3 Dataset Metadata*, http://biostat.mc.vanderbilt.edu/wiki/pub/Main/DataSets/titanic3info.txt

[5] *Titanic*, www.openml.org/d/40945

[6] *IBM HR Analytics Employee Attrition & Performance*, www.kaggle.com/pavansubhasht/ibm-hr-analytics-attrition-dataset

[7] *IMDb Datasets*, www.imdb.com/interfaces/

[8] *Continental United States Hurricane Impacts/Landfalls 1851-2018*, www.aoml.noaa.gov/hrd/hurdat/All_U.S._Hurricanes.html

[9] *The National UFO Reporting Center*, www.nuforc.org/

[10] *UFO Sightings*, www.kaggle.com/NUFORC/ufo-sightings

[11] *Kaggle*, www.kaggle.com/docs/datasets

Cleaning Your Data

Garbage in, garbage out.

—George Fuechsel [1]

Okay, so you've read this book, learned how to create *good data* for your projects, and now all your new datasets are squeaky clean and ready for analysis! However, your research colleague borrowed the book but didn't read it and now has a collection of messy datasets. Now what? Well, next we'll learn about some methods for *detecting bad data* and for *cleaning it up*, often referred to as a component of "data munging" or "data wrangling."

The goal of data cleaning is to get the data *ready* for analysis. But how will you know when it is ready? You need some way to measure or characterize the *quality* of your data. And a way to iteratively apply cleanup methods until a desired (or acceptable) level of quality/readiness is achieved.

Data Cleaning Challenges

There are many challenges to achieving data quality. The three most common are

- *Human error* in the data collection and recording processes

- Addressing *missing data*

- Understanding the sources and meanings of *outliers*

These can be worked on in several ways:

- Visually: Direct *human inspection* of the data to discover errors by viewing with text editors, spreadsheets, or other basic tools

- Programmatically: Coding and running your own error-discovery and correction *programs* using R/Python custom code and libraries or using operating system utilities

- Automated: Running self-prepared or commercial *AI/ML* error-discovery and correction systems

Human errors in collecting and recording measurements and observations are almost inevitable. Errors can be the result of inattention, haste, poorly understood instructions, or bad collection methods. Careful discipline and management of this phase of research can help reduce such errors, especially when collection and recording is not performed directly by the researcher. Data-capture programs that check for entry errors (bounds checks, format verification, etc.) would be ideal for error prevention, *if* you had planned ahead.

There are several ways to inspect and correct dataset problems *after* collection – first, a thorough *manual* (visual) review of the dataset, assuming it is not too large or complex for review. But a few hundred records with a dozen or so variables might be the upper limit for effective direct examination of data for mistakes. It's hard to spot typos like transpositions ("LA" for "AL", for example, or a "23" that should have been "32"). Such inspections should be done by *multiple reviewers* who understand what they are looking for. And of course, the data values should be checked that they conform to their specifications in the metadata. Manual inspection should be an iterative process, repeated as needed until the number of errors found falls below an acceptable level. But as you can see, this can be yet another major time burden prior to analysis.

You can write your own data cleaning, transformation, range checking, and other error-discovery programs in R, Python, Excel, and even using UNIX/Linux/OS X utilities. Built-in functions or add-on libraries for these tools can make this task easier and less time-consuming.

Assessing Data Quality

We discussed some of the causes of "bad data" back in Chapter 1. Since your goal is producing *good data*, how can you detect and measure data quality? Because you need to know when your data is ready for analysis. And if it's not, you need to know how much more work is needed to get it ready.

When you start inspecting data for problems, you can start with counting the number or percent of human/recording errors and the same for missing data with the goal of reducing those occurrences. Recording errors can include nonsensical or contradictory values, such as four-digit postal Zip Codes or impossible geospatial coordinates.

Visual exploration is key; basic displays of each data item's distribution – boxplots and histograms for numeric data and bar charts for categorical – will quickly give you a good first impression about ranges, extremes (outliers), and other unusual distribution characteristics. Then you need to deal with any potential problems you find.

Software and Methods for Data Cleaning

Depending on your computing environment, on your level of programming expertise, and on your data readiness requirements, you have a wide variety of software and methods to choose from for cleaning and transforming your data. The most common such tools include Microsoft Excel, R, Python, and operating system utilities. Some are suitable only for small datasets, others applicable to big data analytics. We'll review a few of these for you.

General Procedures

First, don't start cleaning up your only copy of the dataset! *Make a duplicate*, and work on that one. *Keep a record/log of the changes you make.* Obvious? Yes, but you would be surprised at how many researchers mess up their original data trying to "improve" it. Do some basic data exploration; get a sense of how bad the problems might be. Check the data values against their definitions in the metadata file (you *do* have one of those, don't you?). Ensure you have and understand workable data item names. Check the CSV delimiters and ensure your data values don't embed the delimiter itself. If you didn't number your observations, you should *add a record number* to the dataset so that you can easily find problem records using that number.

Then *make a plan*; specify, conduct, and record each type of cleanup task, like the three main ones (entry error correction, missing value replacements, and outlier resolution). And we *did* warn you that *this process takes time*, maybe lots of it! – which is why we advised creating good data at the outset.

Next we'll highlight just a few of the data cleaning features of common analytics software to illustrate the variety of tools and methods available. They'll help you to find and correct recording errors, to replace (impute) missing data values, and to discover anomalies (outliers). Check the references at the end of the chapter for more sources on how to learn about and use these features.

Microsoft Excel

Excel is widely available on both Windows and macOS, well-documented, and easy to learn/use/understand, but not so good for extremely large datasets (more than one million records). It can easily import and export the commonly used .csv data file format, but be sure to check the exported data item (column) names for usability (no embedded spaces or special characters). Excel will throw errors when attempting to import poorly structured data files.

Excel has an extensive library of data manipulation functions [3], including, to name just a few:

- CLEAN: For removing non-printable characters from string data

- LOWER/UPPER/STRCONV: Converts the case of string data

- SUBSTITUTE: Replaces string data with different data

- TRIM: Removes leading and trailing blanks from string data

- VALUE: Converts string data to numeric

- FORMAT_DATES: Converts date strings to required format

- ROUND: Rounds numeric data to desired precision

- SORT: Sorts data and allows inspection of extreme values

You can select a row or column and then apply one or more of these functions to transform mis-specified data items into a required consisted format. Of course, you can also use Excel for many data summary and analysis tasks. It can calculate means, frequencies, ranges, correlations, and even linear regression models. For details on analytics using Excel, see Apress' *Learn Data Mining Through Excel* [4].

R Project

R is one of the most widely used open source statistical analysis packages by data scientists [5]. And while it is generally used primarily for data exploration, statistical summary calculations and modeling, and visualizations [6], it also has helpful built-in functions and add-on libraries for data cleaning. We'll use two nearly identical sample datasets to illustrate some of these methods: GD-Data.csv [7], a "clean" version, and GD-Data2.csv [8]. The latter is a copy of the former and has some example data cleanup issues to resolve.

Recording Errors

The first indication of a dataset problem can occur when you attempt to load it into an analytical tool such as R. Typical initial load errors include inability to parse the data records into individual data items. This might indicate structural problems with data item names or incorrectly specified delimiters. Then check the assigned data types against their metadata definitions. R is "smart" enough to automatically recognize most computational data types unless there is some violation of a type somewhere in the data. For example, load the sample datasets, inspect their sizes, display a few records, and check the types of several items in each file:

```
> gd <- read.csv("GD-Data.csv", sep=';')
> gd2 <- read.csv("GD-Data2.csv", sep=';')

> dim(gd) # display the dataframe size

[1] 500    8

> head(gd,5)

  gender age degree field wrkfld annsal payfair jobsat
1 Female  40     BS   Engr    Yes     78       4      4
2   Male  39     MS   Engr    Yes     64       4      4
3   Male  36     MS   Comp     No     70       3      4
4   Male  42     MS   Comp    Yes     NA       5      3
5 NotSay  39     BS   Comp    Yes     71       5      3

> class(gd$age)   # check the age data type
[1] "integer"
```

```
> class(gd2$age) # check the age data type in GD-Data2.csv
[1] "factor"
>
```

Something is wrong in file GD-Data2.csv. The data type should be integer. Most likely there's some non-numeric data in that column, perhaps a letter instead of a number. Try to find it visually, and you'll probably miss it.

R has a variety of search functions, including the UNIX-like grep( ). Assuming the error is an uppercase letter (very common), use

```
> errorline = grep('[A-Z]', gd2$age)
> print(errorline)
[1] 13
> print(gd2$age[errorline])
[1] 30
```

Aha! Someone used the letter "O" for a "0" (zero) in record 13. In a similar way, check the assigned types of the other items in your dataset. Then explore the data further, using *frequency counts* for nominal data, for example:

```
> summary(gd2$gender)

female Female   Male NotSay  Other
     4    184    285     13     14
>
> summary(gd2$jobsat)
   1    2    3    4    5    X
  48  127  118  128   78    1
```

This also reveals some miscoded data for gender (lowercase letter for gender value) and for jobsat (improper character code for Job Satisfaction).

Missing Data

The R summary( ) function can tell you about missing data (which R displays as NA, that is "Not Available"):

```
> summary(gd2$annsal)

   Min. 1st Qu.  Median    Mean 3rd Qu.    Max.     NA's
  55.00   73.00   89.00   87.91  101.50  203.50        3
>
```

Also, R's basic is.na( ) function will check and list whether data values are missing (records 4, 65, and 67's annsal are missing):

```
> is.na(gd2$annsal)
...
 [1] FALSE FALSE FALSE  TRUE FALSE FALSE FALSE FALSE FALSE FALSE FALSE FALSE
[13] FALSE FALSE FALSE FALSE FALSE FALSE FALSE FALSE FALSE FALSE FALSE FALSE
[25] FALSE FALSE FALSE FALSE FALSE FALSE FALSE FALSE FALSE FALSE FALSE FALSE
[37] FALSE FALSE FALSE FALSE FALSE FALSE FALSE FALSE FALSE FALSE FALSE FALSE
[49] FALSE FALSE FALSE FALSE FALSE FALSE FALSE FALSE FALSE FALSE FALSE FALSE
[61] FALSE FALSE FALSE FALSE  TRUE FALSE  TRUE FALSE FALSE FALSE FALSE FALSE
[73] FALSE FALSE FALSE FALSE FALSE FALSE FALSE FALSE FALSE FALSE FALSE FALSE
...
```

There are several ways to deal with missing data. You can simply *ignore* (remove) data with missing item values when computing statistics:

```
> mean(gd2$annsal, na.rm = TRUE)
[1] 87.90521
```

But then you might be throwing away good record values for other items. Before we get further into what to do about missing data, it's helpful to understand the different kinds of missing values, classified by the *reason* for their absence:

- MCAR (missing completely at random): Random absence unrelated to other items in the record

- MAR (missing at random): Random absence related to other items in the record (e.g., hourly wage missing and annual income missing)

- MNAR (missing not at random): Absence due to dependence on another item (e.g., hourly wage missing due to job status unemployed)

When you discover missing data in your dataset, you need to determine *why* it is missing based on these classifications, since that helps you to determine what to do with it: ignore it, remove it, or change it. And that will also depend on how many of the data item values are missing. If your dataset (sample) is small, removing records is probably not ideal.

Imputation

One typical solution is to *replace* missing data with some "reasonable" value, a process known as *imputation*. You can choose any of the following methods for replacement:

- Fixed value: Replace any missing data with a constant value, such as 0 (zero).

- Random value: Replace the missing data with a random value from the range of possible values.

- Estimation (from distribution characteristics): Replace a missing nominal or ordinal value with a mode or median, or replace a numeric value with a mean.

- Prediction: Use a regression model or interpolation to estimate what the missing value "should be".

The method you choose will depend on the type of data and on the analysis you need to do as well as on understanding the *effect* of that choice on the analytical results. For example, replacing missing numeric data with a fixed value will affect its distribution (mean and variability). The R Project has extensive guidance on how to decide on what methods to use for imputation [9].

Outliers

An outlier is a data value that is "unusual" in some manner, extremely large perhaps, like an age of 75 in a group of high school graduates. Outliers should not be rejected or corrected without determining if they are errors or are valid data. One easy way to spot outliers is to *visualize* the data item's distribution, for example:

```
> summary(gd2$annsal)
```

Min.	1st Qu.	Median	Mean	3rd Qu.	Max.	NA's
55.00	73.00	89.00	87.91	101.50	203.50	3

```
>
> boxplot(gd2$annsal)
```

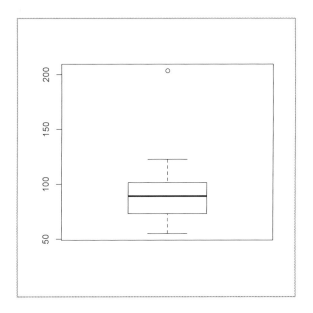

```
> hist(gd2$annsal)
```

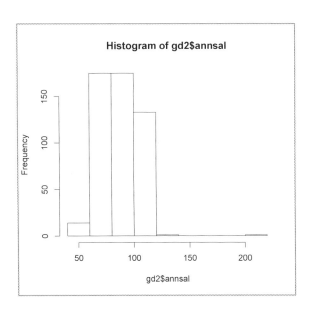

```
> head(sort(gd2$annsal),10)
```

```
 [1] 55.0 56.0 56.0 56.5 56.5 57.0 57.5 58.5 58.5 59.5
> tail(sort(gd2$annsal),10)
 [1] 116.0 116.0 116.7 117.2 117.7 117.7 118.2 119.2 122.7 203.5
```

In this sample dataset, one value for annsal seems unusually high, although not impossible. It requires further investigation. Such outliers can have significant effect on predictive models such as regression, so it's important to understand their validity.

In addition to simple summary statistics and basic visualization graphics which are suitable for smaller datasets, there are also automated AI-based outlier detection methods for large ("big data") datasets [10,11]. These tools use statistical estimation theory to characterize distributions of data items and to report deviations from expected values. Even these methods must be used with care, however, since outliers can represent real, meaningful observations rather than recording errors.

Python

Although R is extensively used for statistical and visual data exploration and analysis [12], Python has similar (some argue better) functionality, and in addition has numerous built-in functions and add-in libraries for data cleaning. And as we saw with R, when Python loads data, it also makes assumptions about the data types it's reading. This too can be used to assist with data cleanup. Reading the example CSV files into two Python *dataframes* uses the pandas library [13]:

```
import pandas as pd
```

```
gd = pd.read_csv('GD-Data.csv', sep=';')
gd2 = pd.read_csv('GD-Data2.csv', sep=';')
```

Using the same sample data files, check the data types Python assumed when it loaded the data:

```
gd.info( )
```

```
<class 'pandas.core.frame.DataFrame'>
RangeIndex: 500 entries, 0 to 499
Data columns (total 8 columns):
gender     500 non-null object
age        500 non-null int64
degree     500 non-null object
field      500 non-null object
wrkfld     500 non-null object
annsal     500 non-null float64
payfair    500 non-null int64
jobsat     500 non-null int64
dtypes: float64(1), int64(3), object(4)
```

Now check the GD-Data2.csv file:

```
gd2.info( )
```

```
<class 'pandas.core.frame.DataFrame'>
RangeIndex: 500 entries, 0 to 499
Data columns (total 8 columns):
gender     500 non-null object
age        500 non-null object
degree     500 non-null object
field      500 non-null object
wrkfld     500 non-null object
annsal     497 non-null float64
payfair    500 non-null int64
jobsat     500 non-null object
dtypes: float64(1), int64(1), object(6)
```

Like R, Python ran into the non-numeric error in the age data and misrepresented its data type. So, you can get the same type of data discovery regardless of your preferred analytical tool. Continuing, if you try to get summary stats on age in GD-Data2.csv, you'll get *frequency counts* instead of the expected mean and percentile stats. Such unexpected results are indications of data errors that must be located and corrected. Visualization can help as we saw using R:

```
gd2.boxplot('annsal')
```

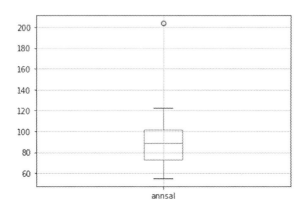

```
gd2.hist('annsal')
```

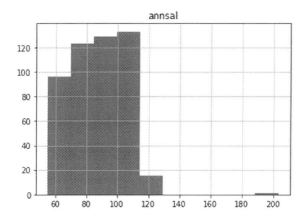

And it's the same process for nominal data like gender:

```
gd2.gender.value_counts()
```

```
Out[13]:
Male      285
Female    184
Other      14
```

```
NotSay      13
female       4
```

```
gd2.gender.value_counts().plot(kind='bar')
```

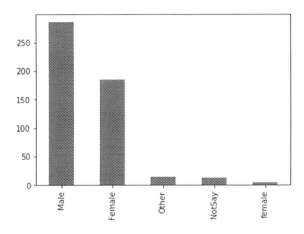

Python can also flag records that contain missing data (locating the same three records found in the previous R example):

```
gd2.isnull( )
```

	gender	age	degree	field	wrkfld	annsal	payfair	jobsat
0	False	False	False	False	False	False	False	False
1	False	False	False	False	False	False	False	False
2	False	False	False	False	False	False	False	False
3	False	False	False	False	False	**True**	False	False
4	False	False	False	False	False	False	False	False
..	...	...	...	...	...	...	...	...

The pandas library for Python [13,14] has several additional functions for identifying and correcting missing data. The dropna( ) function can be used to exclude records with missing values from a data frame, although you might not want to do that depending on the size and nature of your dataset and the potential effect on its summary statistics. Perhaps a better strategy is to *impute* the missing data, and the fillna( ), replace( ), or interpolate( ) functions can be used to substitute it with a fixed value or with the mean, median, or mode of the good data values, although even this strategy can reduce the usefulness of the data's variance and other dispersion statistics.

Operating System Utilities

If you are relatively fluent in UNIX-like operating systems (Linux or Apple's OS X), you can do a surprising amount of data exploration and cleaning tasks using various built-in utilities, such as

- `grep (-i)`
  - Searches for character strings; -i ignores case.
- `find (-exec)`
  - Searches for files and can execute commands on results.
- `tr`
  - Translates string contents (for changing case or making substitutions).
- `sort (-n)`
  - Sorts data values, -n for numeric.
- `sed`
  - Inline stream editor for programmatic changes.
- `cut`
  - Selects sections of records, characters, or fields.
- `uniq (-c)`
  - After sorting, can count unique occurrences.

These OS commands can be combined in various ways, using the output of one command as input to a subsequent command (called "piping" from one to another). For example

```
cut -f1 -d';' GD-Data2.csv | sort | uniq -c
```

in a single line selects the first field (`gender`) from the dataset file, specifying the semicolon delimiter, sorts the output, and then gives a frequency count of the results, thus revealing the existence of some incorrect coding:

```
 184 Female
 285 Male
  13 NotSay
  14 Other
   4 female
   1 gender
```

For another example

```
grep 'MS' GD-Data2.csv | sort -n -t ';' -k 6 -r
```

will search (grep) for all MS values of degree in the dataset and sort on the sixth (annsal) field, showing the extreme value:

```
Female;21;MS;Biol;Yes;203.5;1;1
Male;42;MS;Biol;Yes;116.0;4;5
Male;40;MS;Comp;Yes;116.0;4;2
Male;29;MS;Comp;Yes;116.0;5;4
Male;42;MS;Biol;Yes;115.0;3;4
Male;36;MS;Comp;Yes;115.0;3;3
Female;41;MS;Chem;Yes;114.5;1;2
Female;38;MS;Biol;Yes;114.5;4;4
...
```

The point of this exercise is to show that there are many alternative methods to exploring and summarizing data, depending on your expertise and creativity, including some that don't require R or Python. Exploit what you know for your data analytics requirements, especially in the early data cleanup phase.

AI/ML-Based Software

Recent developments in big data analytics have evolved to deal with data quality issues, especially regarding the needs of AI-based and machine learning algorithms. As more automated prediction and decision software are used for everything from consumer purchasing modeling to prison sentencing, the need for clean, trustworthy data is critical. Model-based and rule-based software for detecting outliers in massive datasets are used extensively in finance and credit card monitoring to detect fraud. Such software can also quickly check for defined relationships among the data items, like checking proper postal Zip Codes against recorded state and city data, including the ability to make automated corrections.

Open source data cleaning software such as HoloClean [10] uses machine learning to predict and impute missing data values. Commercial data exploration systems such as SAS [15] and SPSS [16] provide data management functions similar to those in R and Python, including imputation methods, along with tutorials and documentation on how to clean their respective specialized datasets. Tableau [17] is a widely used data visualization package that is particularly useful in exploring datasets and identifying incorrect and missing data due to its user-friendly graphical interface.

Fortunately for future researchers, data analytics software is evolving rapidly to include new visual cleanup approaches which incorporate natural language processing and artificial intelligence to make this onerous task more efficient and less time-consuming.

Summary

We have only scratched the surface of all the data cleaning issues and fixes that researchers need in order to create good data. Numerous helpful software tools and methods are available, and you can choose the ones that fit your level of expertise and research needs. Fixing errors, addressing missing data items and records, and checking outliers are the core tasks of data cleaning prior to analysis; investigate the tools and methods summarized in this chapter, study the references, and use them when you need to get your data ready.

In Chapter 9, we'll review the endpoint of the research process – doing *good data analytics* – its goals and processes. With good data, you can ensure that your results will be as accurate and trustworthy as possible.

Chapter References

[1] *George Fuechsel*, https://searchsoftwarequality.techtarget.com/definition/garbage-in-garbage-out

[2] *Excel Functions*, www.techonthenet.com/excel/formulas/index.php

[3] *Learn Data Mining Through Excel,* www.apress.com/gp/book/9781484259818

[4] *The R Project for Statistical Computing,* www.r-project.org/

[5] *Beginning R 4,* www.apress.com/gp/book/9781484260524

[6,7] *GD-Data.csv, GD-Data2.csv,* https://github.com/hjfphd/CreatingGoodData

[8] *CRAN Task View: Missing Data,* https://cran.r-project.org/web/views/MissingData.html

[9] *HoloClean,* www.holoclean.io/

[10] *Data Cleaning,* https://dl.acm.org/doi/book/10.1145/3310205

[11] *Python Data Analytics,* www.apress.com/gp/book/9781484239124

[12] *Pandas,* https://pandas.pydata.org/

[13] *Thinking in Pandas,* www.apress.com/gp/book/9781484258385

[14] *SAS Analytics Software,* www.sas.com/en_us/home.html

[15] *IBM SPSS Statistics,* www.ibm.com/products/spss-statistics

CHAPTER 9

Good Data Analytics

Knowledge is the capacity to act.

—Nico Stehr [1]

You've completed your research activities, collected (and cleaned) your dataset, and you're ready to figure out what all that data means. Congratulations, you've been promoted to data analyst! Now your job is to extract *actionable knowledge* from it, for you and for your readers/clients, from that massive pile of numbers and symbols. But now is not the time to just get started thinking about your analysis. As we have emphasized in the previous chapters, you need to plan your analysis requirements long before collecting your data, with the goal of *starting with good data*.

What Is Good Analytics?

As we have emphasized in the previous chapters, *good analytics* starts with an idea, a plan, anticipating the required analyses, and deciding what to collect and how to collect it in order to support that process. Remember the one you've been following:

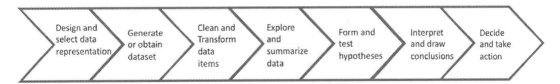

You're now in the latter phases of that sequence, with the ultimate goal of *extracting knowledge* from your data. And that also means your goal is to *use* that knowledge to *make decisions* (or enable others to make them) and *act upon them*. So, your analysis must be transparent, well-defined, well-executed, documented, and reproducible. Assuming you have planned it based on the needs of your original idea and hypotheses,

H. J. Foxwell, *Creating Good Data*, https://doi.org/10.1007/978-1-4842-6103-3_9

you will need to present the statistics and visualizations that support (or reject) those hypotheses.

Although you might want to settle on one favorite analytical tool, be sure to explore the capabilities of alternate software. R is often preferred for statistical summaries and modeling, while Python is more general and helpful with data transformations and specialized data analysis libraries. Both can handle common dataset formats and large file sizes. And there are numerous helpful publications for both languages (already referenced).

Statistical tables and charts need to be readable, uncluttered, and properly labeled (titles, axes, units, legends) and captioned. You should recall our earlier discussions of analytical data types and choose appropriate visualizations for your data – basic frequency counts and bar charts for nominal and ordinal data; variation, central tendency stats, and histograms for ratio data; and crosstabs and scatterplots for showing relationships between pairs of data items. Explore and prepare simple graphics before creating more complex displays (stacked bars, bubble charts, etc.). And if you show multiple adjacent graphs, ensure their vertical axes match.

But by themselves, such displays don't present *knowledge*; that comes from your *interpretation* and *domain expertise*.

When making predictions and extrapolations, remember: your conclusions should be as accurate as possible, relevant, and based on the data.

> *All models are wrong, some are useful.*
>
> —*George Box [2]*

Beware of assigning causality based on correlation; don't overgeneralize from your sample to the whole population, and as we learned in Chapter 6, this is another place to be vigilant against bias, so list your assumptions and potential biases. One of the most important sections of your research report (or data analysis) is an honest and open evaluation of the limitations of your study.

Finally, prepare to share your data and findings, encourage peer review, and respond humbly to corrections and suggestions.

Big Data Analytics

Big data is different. As a reminder, "big data" refers to massive datasets or streams that are extreme in size (Volume), arrival frequency (Velocity), or complexity (Variety) – the typical "V" characteristics used to characterize them. Analyzing such data often requires higher performance systems than the average laptop or perhaps needs cloud-based resources such as Amazon Web Services (AWS) [3]. Additionally, analytical tools and methods must be able to handle these extremes, including machine learning algorithms, NoSQL databases, and specialized languages. We don't cover this aspect of data analytics in this book; we invite you to explore this topic in the references listed here: `www.apress.com/gp/search?query=python+big+data`.

However, our general guidance about data exploration, summary statistics, and visualizations *still applies* to the analysis of such data, with the added warnings that it's more likely to find spurious correlations in large datasets and that statistical significances in hypothesis tests don't necessarily translate into practical or meaningful differences. Additionally, many big data sources are not explicitly designed as data collection processes, they're used for other purposes like social media communities (e.g., Twitter), and so their metadata might not conform to the practices we've covered in this book and might also have many missing data values and other cleanup issues.

Data for Good

In this book, we have focused on "good" data with respect to quality and usability characteristics. But there are other meanings of that word – "beneficial," "trustworthy," or "ethical" – that need further discussion. In many legal systems, testimonies and oaths require "the truth, the whole truth, and nothing but the truth." Applied to research analytics, we should offer "the data, the whole data, and nothing but the data," meaning don't hide or delete inconvenient data nor add false data to research results.

There is a growing concern among both data science practitioners and consumers about the power and influence of modern analytics and the consequences of misuse, and there are many recent examples of this [4]. The data that you gather, analyze, and publish can be used for good or for ill, so you should remember that data gathered on human subjects must be protected and used ethically. For such research, remember that *data are people*. All university- and government-sponsored research on human subjects, from surveys to medical data collection, typically require informed consent, anonymity and/or privacy, and guarantees of safety.

Professional organizations of social and scientific researchers provide guidance for their members on how to conduct human studies, for example, the American Statistical Association's Code of Ethics [5], which includes, in part,

- The ethical statistician uses methodology and data that are relevant and appropriate, without favoritism or prejudice, and in a manner intended to produce valid, interpretable, and reproducible results.

- The ethical statistician is candid about any known or suspected limitations, defects, or biases in the data that may affect the integrity or reliability of the statistical analysis.

- The ethical statistician supports valid inferences, transparency, and good science in general, keeping the interests of the public, funder, client, or customer in mind.

The Data Science Association's Code of Conduct [6] goes further in defining analytics terms along with examples of fraud and misuse of statistics and adds a requirement of *competence* for professional data scientists:

A data scientist shall provide competent data science professional services to a client. Competent data science professional services requires the knowledge, skill, thoroughness and preparation reasonably necessary for the services.

In other words, you need to know what you are doing as a developing professional in this rapidly evolving field, continually learning and increasing your skills and credentials.

These principles ought to apply not only to "statisticians" but to *all researchers* who gather and interpret data. And if you violate them, you are likely to end up like the *discredited* researcher who not only falsified his data but also created great confusion and harm as a consequence [7].

Summary

We have emphasized the need for competent and ethical data analyses in this final chapter, reviewing the data creation and exploration processes required to conduct successful and meaningful research. Use this book along with all the included references to further enhance your understanding and competence as a data science student or practitioner.

The "Recommended Reading" section following this chapter presents publications and resources that should be part of your professional library.

Chapter References

[1] *Nico Stehr*, www.amazon.com/Power-Scientific-Knowledge-Research-Public/dp/1107606721/

[2] *George E. P. Box*, https://en.wikiquote.org/wiki/George_E._P._Box

[3] *Amazon Web Services*, https://aws.amazon.com/

[4] *Data Versus Democracy*, www.apress.com/gp/book/9781484245392

[5] *Ethical Guidelines for Statistical Practice*, www.amstat.org/ASA/Your-Career/Ethical-Guidelines-for-Statistical-Practice.aspx

[6] *DATA SCIENCE CODE OF PROFESSIONAL CONDUCT*, www.datascienceassn.org/code-of-conduct.html

[7] *The MMR vaccine and autism: Sensation, refutation, retraction, and fraud*, www.ncbi.nlm.nih.gov/pmc/articles/PMC3136032/

APPENDIX A

Recommended Reading

Advice on data creation, cleaning, and presentation is scattered among many diverse references from multiple disciplines. *Creating Good Data* consolidates much of this advice. The following references direct readers to further details and guidance on data and measurement practices and theory. If you are a practicing data analyst or are just entering this career, you should be familiar with and have these resources in your professional library.

Books

The Practice of Social Research, 14th ed., E. Babbie, 2015, `www.amazon.com/Practice-Social-Research-Earl-Babbie/dp/1305104943/`

- Don't be put off by the title reference to "Social" research. This well-regarded textbook is applicable to nearly all domains of scientific data collection and analysis. It includes valuable advice on experimental design, selection of statistical methods, and ways to avoid errors and bias in research.

The Visual Display of Quantitative Information, 2nd ed., E. Tufte, 2001, `www.amazon.com/Visual-Display-Quantitative-Information-dp-1930824130/dp/1930824130/`

- Tufte is the undisputed guru of data visualization practices, with four widely praised volumes on turning numbers and concepts into clear and informative pictures. This book introduces his principles of data display and the concepts for clear graphical communication.

© Harry J. Foxwell 2020
H. J. Foxwell, *Creating Good Data*, https://doi.org/10.1007/978-1-4842-6103-3

Python Data Analytics, F. Nelli, 2018, `www.apress.com/us/book/9781484239124`

- Python is one of the premier programming languages for data exploration. It has numerous libraries specifically created for importing and analyzing datasets, such as *pandas*, *numpy*, and *matplotlib*, all essential for producing statistical summaries and visualizations of the data.

Beginning Data Science in R, T. Mailund, 2017, `www.apress.com/us/book/9781484226704`

- Should data scientists learn R or Python? In spite of public debates about this question, industry surveys of educators and practitioners clearly favor both, for statistical analysis, visualization, and data cleaning.

Data Science, J. Kelleher and B. Tierney, 2018, `https://mitpress.mit.edu/books/data-science`

- What exactly is *data science*, and why is it important? This short yet informative booklet includes a bit of history, philosophy, principles, and ethics with respect to the collection and use of data.

Data Cleaning, I. Ilyas and X. Chu, 2019, `http://books.acm.org/titles#tab210`

- A more formal and algorithmic treatment of data cleaning, focusing on automated error detection and repair, and programmatic outlier detection, with an emphasis on machine learning models.

Websites

- Data Science Central
  - `www.datasciencecentral.com/`
  - A comprehensive online resource for literally everything about data science: concepts, books, careers, new methods, collaboration, and past and future. Essential reading; sign up and subscribe to the newsletter.

- Data Science Association

    - www.datascienceassn.org/

    - A non-profit professional association for data scientists. Publishes information and news on DS events and conferences, professional standards, and training for practitioners and students.

- Harvard Data Science Review

    - https://hdsr.mitpress.mit.edu/

    - An open access journal of research and news about data science.

- Carl Sagan's Baloney Detection Kit

    - www.brainpickings.org/2014/01/03/baloney-detection-kit-carl-sagan/

    - Essential reading for a scientific approach to evaluating scientific and non-scientific claims.

Oldies but Goodies

If you can find them! Often reading about the origins of modern technologies helps you understand their importance and basic concepts.

Exploratory Data Analysis, J. Tukey, 1977, Pearson, www.amazon.com/Exploratory-Data-Analysis-John-Tukey/dp/0201076160/

- John Tukey was a famous and frequently cited statistician who developed many of the analytical tools used today in data science, before the widespread use of computers.

A Guide for Selecting Statistical Techniques for Analyzing Social Science Data, 2nd ed., F. Andrews et al., 1981, Survey Research Center, University of Michigan

- This guide provides a valuable flowchart for identifying the four major data types and then selecting appropriate statistical summaries and hypothesis tests.

Selecting Statistical Techniques for Social Science Data: A Guide for SAS, 1st ed., K. Welch, 2008, SAS Institute, `www.amazon.com/Selecting-Statistical-Techniques-Social-Science/dp/1580251188/`

- An update to the Andrews guide, this version includes instructions on using that guide with the SAS statistical software package.

How to Lie with Statistics, D. Huff, 1993, W. W. Norton & Company, `www.amazon.com/How-Lie-Statistics-Darrell-Huff/dp/0393310728/`

- Still in print, Huff's concise guide to avoiding misuse of statistical summaries and visualization.

Applications, Basics, and Computing of Exploratory Data Analysis, P. Velleman & D. Hoaglin, 1981, Wadsworth, `https://ecommons.cornell.edu/bitstream/handle/1813/78/A-B-C_of_EDA_040127.pdf`

- The foundational text on quick methods for exploring data, introducing many of today's commonly used tabulation and visualization methods.

The Elements of Graphing Data, 2nd ed., W. S. Cleveland, 1994, Hobart Press, `https://openlibrary.org/works/OL11364533W/The_elements_of_graphing_data`

- Along with Tufte, Cleveland's guidance on the theory and practice of statistical graphics is essential for researchers who must effectively communicate their results visually.

Index

A

Amazon Web Services (AWS), 95
Analytical data types, 12
    aspiring data, 43
    commercial products, 44
    describe, compare/predict, 12
    diastolic and systolic data, 38
    interval data, 23–24
    media data, 25
    nominal/categorical data, 13–16
    ordinal data, 16–21
    questions/methods, 38
    ratio data, 21–23
    sentiment analysis, 24
    text data, 24–25
    visualization, 39–43
Analytic data types
    data process, 93
    extracting knowledge, 93–94
    interpretation/domain expertise, 94
    statistical tables/charts, 94

B

Bias
    availability/convenience bias, 63
    cognitive bias, 60
    collection issues, 62–64
    confirmation bias, 61
    consequences, 65

    data collection bias, 60
    expectancy bias, 63
    goals, 59
    homogeneous sampling, 64
    measurement bias, 63
    outliers, 64
    quota sampling, 63
    recognizing/reducing bias, 64–65
    sampling, 61–62
    selection bias, 60
    temporal bias, 62
    unbiased research data, 62
Big data analytics, 94–95

C

Categorical data, *see* Nominal
    data types
Cleaning data, 75
    AI/ML based software, 89–90
    data quality, 75
    detect/measure data quality, 77
    Excel sheet, 78
    human errors, 76
    inspect/correct dataset, 76
    operating systems utilities, 88–89
    procedures, 77
    Python dataframes, 84–89
    R project, 79–84
    visual exploration, 77

© Harry J. Foxwell 2020
H. J. Foxwell, *Creating Good Data*, https://doi.org/10.1007/978-1-4842-6103-3